Springer Theses

Recognizing Outstanding Ph.D. Research

Aims and Scope

The series "Springer Theses" brings together a selection of the very best Ph.D. theses from around the world and across the physical sciences. Nominated and endorsed by two recognized specialists, each published volume has been selected for its scientific excellence and the high impact of its contents for the pertinent field of research. For greater accessibility to non-specialists, the published versions include an extended introduction, as well as a foreword by the student's supervisor explaining the special relevance of the work for the field. As a whole, the series will provide a valuable resource both for newcomers to the research fields described, and for other scientists seeking detailed background information on special questions. Finally, it provides an accredited documentation of the valuable contributions made by today's younger generation of scientists.

Theses are accepted into the series by invited nomination only and must fulfill all of the following criteria

- They must be written in good English.
- The topic should fall within the confines of Chemistry, Physics, Earth Sciences, Engineering and related interdisciplinary fields such as Materials, Nanoscience, Chemical Engineering, Complex Systems and Biophysics.
- The work reported in the thesis must represent a significant scientific advance.
- If the thesis includes previously published material, permission to reproduce this must be gained from the respective copyright holder.
- They must have been examined and passed during the 12 months prior to nomination.
- Each thesis should include a foreword by the supervisor outlining the significance of its content.
- The theses should have a clearly defined structure including an introduction accessible to scientists not expert in that particular field.

More information about this series at http://www.springer.com/series/8790

Dale Jonathan Waterhouse

Novel Optical Endoscopes for Early Cancer Diagnosis and Therapy

Doctoral Thesis accepted by
the University of Cambridge, Cambridge, UK

Author
Dr. Dale Jonathan Waterhouse
Department of Physics and Cancer
Research UK Cambridge Institute
University of Cambridge
Cambridge, UK

Supervisor
Dr. Sarah Bohndiek
Department of Physics and Cancer
Research UK Cambridge Institute
University of Cambridge
Cambridge, UK

ISSN 2190-5053 ISSN 2190-5061 (electronic)
Springer Theses
ISBN 978-3-030-21483-8 ISBN 978-3-030-21481-4 (eBook)
https://doi.org/10.1007/978-3-030-21481-4

This Springer imprint is published by the registered company Springer Nature Switzerland AG
The registered company address is: Gewerbestrasse 11, 6330 Cham, Switzerland

To my mum and dad,
whose love gave me the strength to succeed,
and to my brother Aaron,
my best friend
...

Supervisor's Foreword

Optical imaging plays an important role in the early diagnosis of cancer. In particular, the use of optical endoscopes to relay image information from deep within the body to the external observer is widespread. At present, however, the optical information relayed is typically recorded using a standard colour camera, which integrates a Bayer filter array of red, green and blue colour filters to replicate the colour sensing capability of the human eye. This approach restricts the range of wavelengths that can be detected and the number of spectral features that can be resolved. Using standard colour cameras provides limited contrast for cancer within a background of normal tissue, which leads to high miss rates during endoscopic surveillance and difficulty in assessment of tumour margins during surgery.

The interactions of light with tissue go far beyond the simple red, green and blue. In addition to those interactions that occur intrinsically with different biomolecules in the tissue, it is also possible to use contrast agents that can specifically enhance the contrast between healthy and diseased tissues. These contrast agents may be untargeted and give rise to contrast due to differences in tissue structure or vascularisation, or they may be targeted to specific cell surface receptors or other biological processes that are known to change during disease progression. The detection of these contrast agents in tissue can often be difficult due to the high background signals that arise from the intrinsic tissue interactions.

In this thesis, Dale Waterhouse has begun to exploit these additional sources of contrast in a series of clinically motivated studies. The thesis begins with the development of an optical imaging biomarker road map, which critically reviews the opportunities and challenges for optical imaging in endoscopy and identifies common features of successful optical imaging biomarkers that have been deployed clinically. This defined framework is then kept at the forefront for the remainder of the thesis, in which the design, development, characterisation and application of several novel imaging systems are presented.

Dale then embarked upon the development of a bimodal white light and near-infrared endoscope for imaging of a targeted fluorescence contrast agent in the context of early detection of dysplasia in patients with Barrett's oesophagus. With the clinical translation in mind, the endoscope was built around an existing

CE-marked device, modifying only the back end of the system. He then extended the initial system further to resolve intrinsic contrast from different biomolecules by introducing a multispectral imaging capability using a spectrally resolved detector array as the camera within the system. Taking the approach of using a CE-marked imaging fibre bundle for these endoscopy experiments required the development of a number of computational methods to deal with comb artefacts and overcome challenges with demosaicking of images from the multispectral camera. In addition to applying these advanced optical imaging methods in flexible endoscopic imaging, Dale also created and clinically applied a multispectral imaging system that could be used to guide surgery for the removal of pituitary adenomas.

With his dissertation, Dale made an important contribution towards understanding how optical imaging can be applied in biomedicine. In particular, by developing a clear understanding of the characteristics shared by successful optical imaging biomarkers, Dale ensured the success of his own device development, with two of his systems being applied in 'first-in-human' clinical trials. This is a significant achievement within only a 4-year time frame and is a testament to the impact of Dale's thinking not only on the research of my own laboratory, but also on the field of biomedical optical imaging more generally.

Cambridge, UK
June 2019

Dr. Sarah Bohndiek

Abstract

Imaging is the only medical tool currently capable of non-invasively capturing detailed, real-time and spatially resolved biochemical information in vivo and thus delineating disease so that non-invasive curative resection or treatment of the affected area can take place. Though visible and near-infrared (NIR) light undergo a wide range of complex interactions in tissue—interactions which can be harnessed to yield useful information about the underlying pathology—optical imaging has yet to be fully utilised in clinic, with many existing techniques relying on standard colour imaging that replicates human vision.

This thesis describes my recent effort advancing novel optical endoscopic imaging techniques towards the clinical translation. Before embarking on the development of novel devices, an analysis was made of the common challenges in translating optical imaging techniques. Through this work, a streamlined road map to clinical translation was developed, and key translational characteristics were defined. These were used to guide the subsequent development of endoscopic devices.

Initial efforts were focused on the development of flexible endoscopes for detection of dysplasia in Barrett's oesophagus. To enable molecular imaging with a newly discovered targeted fluorescent contrast agent, a bimodal endoscope capable of capturing NIR fluorescence and white light reflectance was developed around a clinically translatable device architecture, and image artefacts were addressed by developing and evaluating image correction algorithms. This technique demonstrated significant potential for delineation of dysplasia in ex vivo samples. Next, a multispectral endoscope capable of imaging multiple fluorophores or endogenous tissue reflectance was developed. This device was successfully translated to a clinical pilot study, where initial results showed the promising potential of multispectral endoscopy for delineation of dysplasia based on endogenous reflectance from oesophageal tissue. Finally, multispectral imaging was explored for intraoperative delineation of adenoma and healthy pituitary tissue. A novel rigid multispectral endoscope was developed, preliminary technical characterisation of this device was performed, and a clinical pilot study was planned.

With the continuation of this work as outlined at the end of this thesis, the novel techniques described here have the potential to improve the standard of care in their respective indications.

List of Publications

Publications

Jonghee Yoon, James Joseph, **Dale J. Waterhouse**, A. Siri Luthman, George S. D. Gordon and Massimiliano di Pietro, Wladyslaw Januszewicz, Rebecca C. Fitzgerald, Sarah E. Bohndiek, *A clinically translatable hyperspectral endoscopy (HySE) system for imaging the gastrointestinal tract.* Nature Communications, 10, 1902, (2019).

Dale J. Waterhouse, Catherine R. M. Fitzpatrick, Brian M. Pogue, James O'Connor and Sarah E. Bohndiek, *A roadmap for the clinical implementation of optical-imaging biomarkers.* Nature Biomedical Engineering, 3, 339–353, (2019).

Dale J. Waterhouse, A. Siri Luthman, Jonghee Yoon, George S. D. Gordon and Sarah E. Bohndiek, *Quantitative evaluation of comb-structure removal methods for multispectral fiberscopic imaging.* Scientific Reports, 8, 17801, (2018).

A. Siri Luthman, **Dale J. Waterhouse**, Laura Ansel-Bollepalli, Jonghee Yoon, George S. D. Gordon, James Joseph, Massimiliano di Pietro, Wladyslaw Januszewicz and Sarah E. Bohndiek, *Bimodal Reflectance and Fluorescence Multispectral Endoscopy based on Spectrally Resolving Detector Arrays.* Journal of Biomedical Optics, 24(3), (2018).

Dale J. Waterhouse, Catherine R. M. Fitzpatrick, Massimiliano di Pietro and Sarah E. Bohndiek, *Emerging Optical Methods for Endoscopic Barrett's Surveillance.* The Lancet Gastroenterology and Hepatology, 3(5), (2018).

André A. Neves, Massimiliano di Pietro, Maria O'Donovan, **Dale J. Waterhouse**, Sarah E. Bohndiek, Kevin M. Brindle and Rebecca C. Fitzgerald, *Detection of early neoplasia in Barrett's esophagus using lectin-based near-infrared imaging: an ex vivo study on human tissue.* Endoscopy, 50(6), (2018).

Dale J. Waterhouse, James Joseph, André A. Neves, Massimiliano di Pietro, Kevin M. Brindle, Rebecca C. Fitzgerald and Sarah E. Bohndiek, *Design and validation of a near-infrared fluorescence endoscope for detection of early esophageal malignancy.* Journal of Biomedical Optics, 21(8), (2016).

Conference Proceedings

A. Siri Luthman, **Dale J. Waterhouse**, Laura Bollepalli, James Joseph and Sarah E. Bohndiek, *A multispectral endoscope based on spectrally resolved detector arrays.* Proc. SPIE 10411, Clinical and Preclinical Optical Diagnostics, 104110A (2017).

Dale J. Waterhouse, A. Siri Luthman and Sarah E. Bohndiek, *Spectral band optimization for multispectral fluorescence imaging.* SPIE BiOS (2017).

Massimiliano di Pietro, André A. Neves, Maria O'Donovan, **Dale J. Waterhouse**, Sarah E. Bohndiek, Kevin M. Brindle and Rebecca C. Fitzgerald, *Detection of dysplasia in Barrett's oesophagus using lectin-based near infra-red molecular imaging: an ex-vivo study on human tissue*. Proceedings of the British Society of Gastroenterology Meeting (2016).

Dale J. Waterhouse, James Joseph, André A. Neves, Massimiliano di Pietro, Kevin M. Brindle, Rebecca C. Fitzgerald and Sarah E. Bohndiek, *Design and validation of a near-infrared fluorescence endoscope for detection of early esophageal malignancy using a targeted imaging probe*. SPIE BiOS (2016).

Acknowledgements

First, I would like to thank my supervisor, Sarah Bohndiek, for giving me the opportunity to join VISIONLab in 2014. I could not have asked for a better supervisor. Throughout my time in her laboratory, Sarah made time to provide scientific advice and feedback and supported me in purchasing equipment, in travelling to conferences and in my professional development. Despite the laboratory expanding rapidly, Sarah's schedule, along with her office door, has remained open. Her openness, efficiency and the apparent ease with which she carries out her work are inspirational, and these qualities have cultivated a similarly efficient yet relaxed atmosphere in VISIONLab. Never have I felt the crippling pressure of my supervisor bearing down upon me, a feeling all too often described by my peers.

Still, experiments failed. Equipment broke. Code crashed. During these times, I am grateful to have been surrounded by supportive colleagues. As a novice to the research, I am thankful that James Joseph and George Gordon were so willing to show me the ropes. I would also like to acknowledge the support of my exceptional clinical collaborators, especially Massimiliano di Pietro, Wladyslaw Januszewicz and James Tysome, whose support and patience have helped facilitate the clinical translation of my work. I am grateful to have worked alongside Siri Luthman, tackling our problems together in an otherwise desolate optics laboratory.

I would also like to thank my colleagues turned close friends. To Isa, James and Judith, for conversations over coffee, in the car and over beers. Your warm words were always reassuring. To Abby, for helping me to keep calm, and for reading and correcting this thesis. To Michal, for all the fun we shared over the years. Further thanks go to my friends in college, especially Andrea, for the relaxing evenings watching trash TV, and Alexis, for the formals and nights out which made the end of the week worth looking forward to. And a huge thank you to Lina, for her love and support in the final stretch.

Sadly, my big nan-nan is not here to see me complete my thesis. She is dearly missed, and I am sure she would be immensely proud, as are my other grandparents. Their pride and love have spurred me on throughout my research. I express immense gratitude to my mum, dad and brother, Aaron. The weekends we spent in

Cambridge, and at home, helped to remind me of the world outside the Cambridge bubble. Even in their absence, their love, support and advice stayed with me every step of the way, strengthening me in challenging times. The lessons my parents have taught me, and the example they set, have shaped all aspects of who I am and where I am today, and I will be forever grateful.

Finally, I would like to acknowledge the funding from the CRUK-EPSRC Cancer Imaging Centre in Cambridge and Manchester, without which I could not have carried out this work. These thanks extend to the millions of people whose generous donations ensure this work continues.

Contents

1 Translation of Optical Imaging Techniques 1
1.1 The Potential of Optical Imaging in Oncology 1
1.2 Challenges of Clinical Translation of Optical Imaging Techniques 5
1.3 The Optical Imaging Biomarker Roadmap 6
1.3.1 Barriers to Translation of Optical Imaging Techniques 8
1.3.2 Translational Characteristics of Optical Imaging Techniques 11
1.4 Outlook 12
References 13

2 Flexible Endoscopy: Early Detection of Dysplasia in Barrett's Oesophagus 17
2.1 Barrett's Oesophagus 17
2.2 Improving the Standard of Care 19
2.2.1 Categories of Advanced Optical Endoscopic Techniques 19
2.2.2 Advanced Optical Endoscopic Techniques in Clinical Practice for Barrett's Surveillance 20
2.2.3 Emerging Optical Endoscopic Techniques for Barrett's Surveillance 27
2.3 Summary 36
References 37

3 Flexible Endoscopy: Device Architecture 43
3.1 Flexible Endoscopic Device Architectures 43
3.2 The PolyScope Accessory Channel Endoscope 48
3.3 Comb Correction Methods 51
3.3.1 Experimental System 53
3.3.2 Simulated Images 53

3.3.3 Monochrome Image Comb Correction 55
3.3.4 Multispectral Image Comb Correction 57
3.3.5 Performance Metrics 59
3.3.6 Overall Performance 62
3.4 Comb Correction Results 63
3.4.1 Performance of Simulated Monochrome Image Corrections 63
3.4.2 Performance of Experimentally Captured Monochrome Image Corrections 64
3.4.3 Performance of Captured Multispectral Image Corrections 65
3.5 Discussion and Conclusions 68
References 72

4 Flexible Endoscopy: Optical Molecular Imaging 75
4.1 Molecular Imaging for Endoscopic Surveillance of Barrett's Oesophagus 75
4.2 The Potential of Fluorescently Labelled Lectins 77
4.3 Challenges in Realising the Potential of WGA-IR800 78
4.4 Materials and Methods 79
4.4.1 Fluorescent Lectin Synthesis 79
4.4.2 Endoscope Design 79
4.4.3 Image Acquisition and Image Corrections 80
4.5 Technical Characterisation 81
4.5.1 Methods 81
4.5.2 Results 83
4.6 Validation Using Biological Samples 87
4.6.1 Methods 88
4.6.2 Results 90
4.7 Discussion and Conclusions 95
References 99

5 Flexible Endoscopy: Multispectral Imaging 101
5.1 Multispectral Fluorescence Imaging of Targeted Fluorescent Molecules 101
5.2 Multispectral Imaging of Endogenous Contrast 101
5.3 Spectrally Resolved Detector Arrays (SRDAs) 102
5.4 Materials and Methods 103
5.4.1 Fluorescent Contrast Agents 103
5.4.2 Endoscope Design 104
5.4.3 Image Acquisition and Image Corrections 105
5.4.4 Spectral Unmixing 105

5.5 Multispectral Fluorescence Imaging (MFI) 107
5.5.1 Methods 107
5.5.2 Results 110
5.6 Multispectral Reflectance Imaging of Endogenous Tissue Contrast 115
5.6.1 Methods 116
5.6.2 Results (MuSE 01) 117
5.7 Discussion and Conclusions 120
References 124

6 Rigid Endoscopy for Intraoperative Imaging of Pituitary Adenoma 127
6.1 Pituitary Adenoma 127
6.1.1 Standard of Care 127
6.1.2 Advanced Imaging of Pituitary Adenoma 128
6.2 Endoscope Design 131
6.3 Technical Characterisation 132
6.3.1 Methods 132
6.3.2 Results 133
6.4 Clinical Trial 135
6.4.1 Methods 135
6.4.2 Results 136
6.5 Conclusions and Future Work 136
References 137

7 Conclusions and Outlook 139
References 142

Abbreviations

4QB	Four quadrant biopsies according to the Seattle protocol
a/LCI	Angle-resolved low-coherence interferometry
AA	Acetic acid
ACG	American College of Gastroenterology
AF647/700	Alexa Fluor 647/700
AFI	Autofluorescence imaging
AGA	American Gastroenterology Association
ASGE	American Society for Gastrointestinal Endoscopy
ASR	Accuracy of spectral reconstruction
AUC	Area under the curve
BLI	Blue laser imaging
BSG	British Society of Gastroenterologists
CARS	Coherent anti-Raman spectroscopy
CCD	Charge-coupled device
CFA	Colour filter array
CMOS	Complementary metal-oxide-semiconductor
CT	Computed tomography
DRS	Diffuse reflection spectroscopy
eCLE	Endoscope-based confocal laser endomicroscopy
EM	Electron multiplying
EMCCD	Electron multiplying charge-coupled device
EMR	Endoscopic mucosal resection
ERCP	Endoscopic retrograde cholangiopancreatography
ESGE	European Society for Gastrointestinal Endoscopy
ESS	Elastic scattering spectroscopy
ETMI	Endoscopic trimodal imaging
FDA	Food and drug administration
FICE	Fujinon intelligent chromoendoscopy
FLIM	Fluorescence lifetime imaging
FOV	Field of view

FWHM	Full width half maximum
GI	Gastrointestinal
GMP	Good manufacturing practice
GOJ	Gastro oesophageal junction
GPU	Graphics processing unit
HD	High definition
HGD	High-grade dysplasia
IB	Imaging biomarker
ICG	Indocyanine green
IMC	Intramucosal carcinoma
iMRI	Intraoperative magnetic resonance imaging
IR800	IRDye 800CW
iSCAN	A brand name used by Pentax to refer to their image enhancement technology
IV	Intravenous
LCTF	Liquid crystal tunable filter
LED	Light-emitting diode
LGD	Low-grade dysplasia
LSS	Light scattering spectroscopy
MB	Methylene blue
MFI	Multispectral fluorescence imaging
MPE	Maximum permissible exposure
MPM	Multi-photon microscopy
MRI	Magnetic resonance imaging
MSI	Multispectral imaging
NA	Numerical aperture
NAD(P)H	Nicotinamide adenine dinucleotide (phosphate)
NBI	Narrow-band imaging
ND	Non-dysplastic
NIR	Near infrared
OAC	Oesophageal adenocarcinoma
OCT	Optical coherence tomography
OIB	Optical imaging biomarker
OMI	Optical molecular imaging
OP	Overall performance
PAE	Photoacoustic endoscopy
PBS	Phosphate-buffered saline
pCLE	Probe-based confocal laser endomicroscopy
PET	Positron emission tomography
QE	Quantum efficiency
RGB	Red green blue
ROI	Region of interest
RS	Raman spectroscopy
SBR	Signal-to-background ratio
SNR	Signal-to-noise ratio

SOP	Standard operating procedure
SRDA	Spectrally resolved detector array
TC	Translational characteristic
UHD	Ultra-high definition
UHP	Ultra-high powered
UTNE	Unsedated transnasal endoscopy
VI	Visual interface
VLE	Volumetric laser endomicroscopy
WD	Working distance
WGA	Wheat germ agglutinin
WL	White light
WLE	White light endoscopy/endoscope

Chapter 1
Translation of Optical Imaging Techniques

Optical imaging has tremendous potential for non-invasive detection and characterisation of diseased tissue using non-ionising radiation with wavelengths from 400–1000 nm. The approach benefits from: real-time analysis of tissue biochemistry based on sensitive interactions of light and tissue; compact, point-of-care and low-cost implementations compared to radiological imaging; and operation across a range of resolutions and depths that span over 4 orders of magnitude [1].

This thesis examines the potential of optical imaging in two key indications: early detection of cancer in the oesophagus and delineation of cancer and healthy tissue in resection of pituitary tumours. Before embarking on the development of novel devices, an analysis was made of the common challenges in translating optical imaging techniques.

1.1 The Potential of Optical Imaging in Oncology

Visible and near infrared (NIR) light undergo a wide range of complex interactions in tissue, including absorption, reflection, elastic and inelastic scattering, and fluorescence absorption and emission [2] (Fig. 1.1). Conventional imaging discards this rich information, instead capturing reflected red, green and blue light to simply replicate human vision [3]. Instead of discarding this information, we can exploit it to yield optical imaging biomarkers (OIBs), defined characteristics measured as indicators of biology, often derived from measurements made on an image (Fig. 1.2) [4]. A large number of OIBs have been developed to assess hallmarks of cancer [5], ranging from microstructural change and collagen cross-linking associated with replicative immortality and activating invasion, to haemoglobin concentration and oxygenation, associated with induction of angiogenesis. The ability to use these biomarkers to generate high contrast for cancer, has led to clinical approval of dozens of optical imaging techniques for early cancer detection (Table 1.1).

D. J. Waterhouse, *Novel Optical Endoscopes for Early Cancer Diagnosis and Therapy*, Springer Theses, https://doi.org/10.1007/978-3-030-21481-4_1

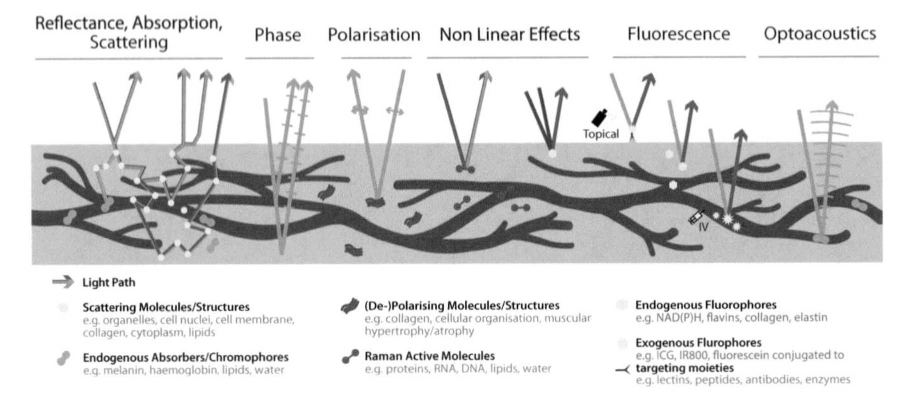

Fig. 1.1 Visible and near-infrared light-tissue interactions. Visible and near-infrared light undergo a wide range of complex interactions with tissue constituents. Novel optical imaging techniques attempt to use these interactions as a source of contrast for clinical imaging. Interactions illustrated include: reflection, absorption and scattering (e.g. white light imaging, diffuse reflectance spectroscopy, narrow band imaging); phase (e.g. optical coherence tomography); polarisation (e.g. polarimetry); non-linear effects (e.g. Raman spectroscopy, multi-photon fluorescence imaging,); fluorescence (e.g. autofluorescence intensity or lifetime imaging of endogenous fluorophores, optical molecular imaging of exogenous fluorophores); and the optoacoustic effect (e.g. photo (or opto) acoustic microscopy or tomography). Line colours represent colours of light. Multi-coloured lines represent broadband light. Changes of colour upon interaction represent change of wavelength, for example, during fluorescence the change from orange to red signifies shorter wavelength absorption and longer wavelength emission. Perpendicular lines represent wave fronts and thus show optical coherence of phase. Arrowed perpendicular lines represent polarisation orientation. Curved lines represent emission of acoustic waves (IV = intravenous)

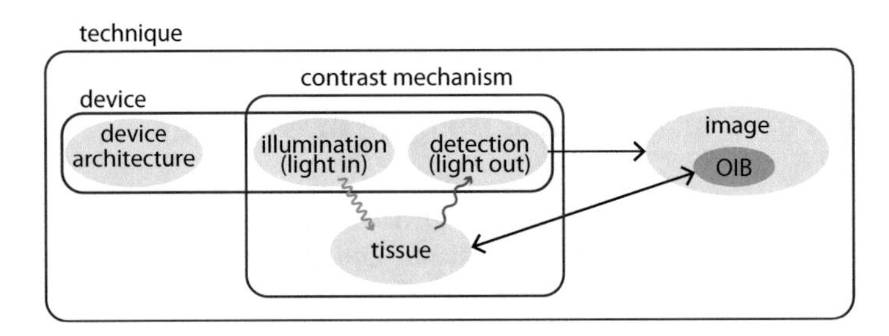

Fig. 1.2 Imaging technique definitions. The terminology surrounding optical imaging is not standardised in the literature. Throughout this thesis we refer to the illumination or 'light in' (e.g. 300 nm light), detection or 'light out' (e.g. 400 nm light), tissue constituent of interest (e.g. collagen) and their light-tissue interaction (e.g. collagen fluorescence) as the contrast mechanism. When the choice of light is combined with a device architecture (e.g. microscope, endoscope), we refer to this as a device. The light out is used to create an image that encodes the underlying light-tissue interaction. Optical imaging biomarkers (OIBs) are defined characteristics of the image. These could be simple, such as image intensity, or more complex, such as degree of branching in vascular pattern. Once properly validated, OIBs can act as indicators of the underlying tissue biology, for example disease pathology. We refer to the combination of device, contrast mechanism, and OIB as a 'technique'. We avoid use of the term 'modality' as it is used variably throughout the literature. Note that these definitions may be different to those used in non-optical imaging

Despite this potential, and the obvious enthusiasm for such technologies to reach clinical implementation, whilst many techniques have been approved for use in humans, relatively few have reached routine use in healthcare systems, begging the question: what is preventing their translation into healthcare?

Table 1.1 Optical endoscopic imaging techniques in development in oncology

Approach	Source of contrast	Biological change in cancer	Example OIB	Clinically approved commercial devices
White light imaging [6]	Endogenous chromophores (e.g. melanin, haemoglobin, tryptophan, lipids, water); endogenous scatterers (e.g. mitochondria, cell nuclei, cell membrane)	Vasculature, melanin distribution, visible lesions	Mucosal abnormalities	Widely available endoscopes and dermoscopes, naked eye examination
Virtual (or electronic) chromoendoscopy [7–11]			Irregular mucosal and vascular patterns	Olympus NBI Scopes [12], Fujicon FICE [13]. PENTAX iScan [14], SIMSYS-MoleMate™ [15], MelaFind® [16], Spectral Molecular Imaging SkinSpect™ [17]
Polarimetry [18]	Microstructural anisotropy	Collagen, cellular orientations, organisation, muscular hypertrophy/atrophy	Unclear	Derma Medical MoleMax 3 [19]
Dye-based Imaging				
Chromoendoscopy [9, 11]	Dyes/stains (e.g. methylene blue, acetic acid). Can enhance surface topology or be differentially absorbed by different cell types	Visible lesions, cell type	Loss of acetowhitening	Widely available endoscopes

(continued)

Table 1.1 (continued)

Approach	Source of contrast	Biological change in cancer	Example OIB	Clinically approved commercial devices
Targeted fluorescence imaging [20, 21]	Exogenous fluorophores conjugated to targeting moieties (lectins, peptides, antibodies, affibodies, enzymes) that target intracellular and extracellular proteins and enzymes	Changes in targeted intracellular and extracellular proteins and enzymes	High intensity fluorescence due to uptake or binding of probe	None
Endoscopic microscopy [9, 22]	Exogenous fluorophores (e.g. fluorescein)	Cell type	Change in cell phenotype such as irregular borders and shapes	Mauna Kea Cellvizio® [23], Optiscan ViewnVivo™ [24], Caliber ID VivaScope [25]
Autofluorescence Imaging				
Autofluorescence intensity imaging [26]	Endogenous fluorophores (e.g. NAD(P)H, flavins, collagen, elastin, phenylalanine, tryptophan)	Biochemical e.g. collagen cross-link breaking	Loss of autofluorescence, denoted by purple image highlight	Widely available endoscopes
Fluorescence Lifetime Imaging [27–29]			Change in melanocyte distribution	JenLab DermaInspect [30]
Multi photon microscopy [27–29]			Change in melanocyte distribution	JenLab DermaInspect [30]
Cross Sectional Imaging				
Optical coherence tomography [31]	Boundaries between structural features	Vasculature, visible lesions, structural changes	Change in structure such as irregular glandular architecture	Ninepoint Medical NvisionVLE® [32], Verisante Core™ [33], Skintell [34], Michelson Diagnostics Vivosight [35]

(continued)

Table 1.1 (continued)

Approach	Source of contrast	Biological change in cancer	Example OIB	Clinically approved commercial devices
Photoacoustic imaging	Absorption of endogenous chromophores (e.g. melanin, haemoglobin)		Reduction in oxygen saturation	None
Spectroscopy				
Raman spectroscopy [36–39]	Vibrational modes of specific molecules (e.g. proteins, DNA, lipids)	Biochemical	Change in spectrum	Verisante Aura™ [40]
Reflection spectroscopy [41, 42]	Reflection spectra of endogenous biological molecules	Biochemical	Change in spectrum	Pentax WavSTAT4™ [14]
Diffuse reflectance spectroscopy [43, 44]	Endogenous scatterers (e.g. mitochondria, cell nuclei, cell membrane, collagen, cytoplasm, lipids)	Vasculature, change in distribution of cells, nuclei and organelles	Change in spectrum	None

FOV: field of view

1.2 Challenges of Clinical Translation of Optical Imaging Techniques

Development of novel optical imaging techniques often begins with the discovery of promising OIBs using optical devices dedicated to ex vivo imaging. This can motivate the development of devices able to measure the same biomarker in vivo (this is our objective in Chap. 4). Alternatively, development may begin when application of a contrast mechanism and/or device architecture found to be useful in another field is applied to a new biological question (this is our objective in Chap. 6).

For any new imaging biomarker to be deployed in healthcare, detailed technical validation is required to define the precision and accuracy with which the biomarker can be measured. Biological validation, establishing the association between the biomarker and the underlying physiological, anatomical, or pathological process, should be performed in parallel with technical validation. Importantly, clinical validation, establishing that the biomarker identifies, measures, or predicts the outcome of interest, can be performed only with an imaging device that is approved for use in patients.

Considering these validation processes, there are two main differences in the clinical translation of a novel OIB compared to an imaging biomarker developed for standard radiological imaging, such as computed tomography (CT) or magnetic resonance imaging (MRI), where the imaging device is already clinically approved for use in humans and widely available in hospital radiology departments [4]. Firstly, for OIBs, it is uncommon that an existing clinically approved imaging device, with associated specialist data acquisition and interpretation methods, will already be available to allow clinical validation of the biomarker. As a result, biological validation may be restricted to testing with ex vivo samples, such as histopathological sections, which can be prone to bias and generate a different range of optical interactions compared to the in vivo setting; the result can be misleading conclusions as to the potential clinical utility of the OIB [9, 45].

A second key difference between optical and radiological imaging biomarkers is that OIBs may be deployed in a range of settings in the clinical patient management pathway, spanning the home; primary care (family physician); and specialist care (specialist practice, referral or medical centre). Thus, even a well-defined biomarker with promising performance in the experimental setting may never receive approval if it fails to adequately address a specific diagnostic question at the appropriate point in the care pathway.

1.3 The Optical Imaging Biomarker Roadmap

With the differences between radiological imaging biomarkers (IBs) and OIBs in mind, we developed a specific 'Optical Imaging Biomarker Roadmap' (Fig. 1.3) based on the international consensus 'Imaging Biomarker Roadmap' created for use in cancer studies [4]. The aim of the roadmap is to elucidate the reasons for failure in clinical translation of optical imaging techniques and to overcome these limitations to allow smoother translation of promising techniques into clinic.

In all domains, optical device development plays a larger role in translation than in standard radiological imaging, where a device to measure the imaging biomarker is usually already clinically approved. Throughout the OIB Roadmap, there is a complex interplay of limitations imposed by the device, the contrast mechanism and the OIB, meaning technical and biological validation cannot be considered in isolation. For example, the precision (repeatability/reproducibility) of measuring a given OIB defined with respect to a perfect test target will differ considerably from precision defined with respect to the biological measurement made in a patient, which is directly relevant in the clinical application. Common feedback loops that often inhibit translation are highlighted so that advance consideration might allow developers to avoid them. Construction of the roadmap revealed five key areas where experimental methodologies or regulations can create translational barriers. These

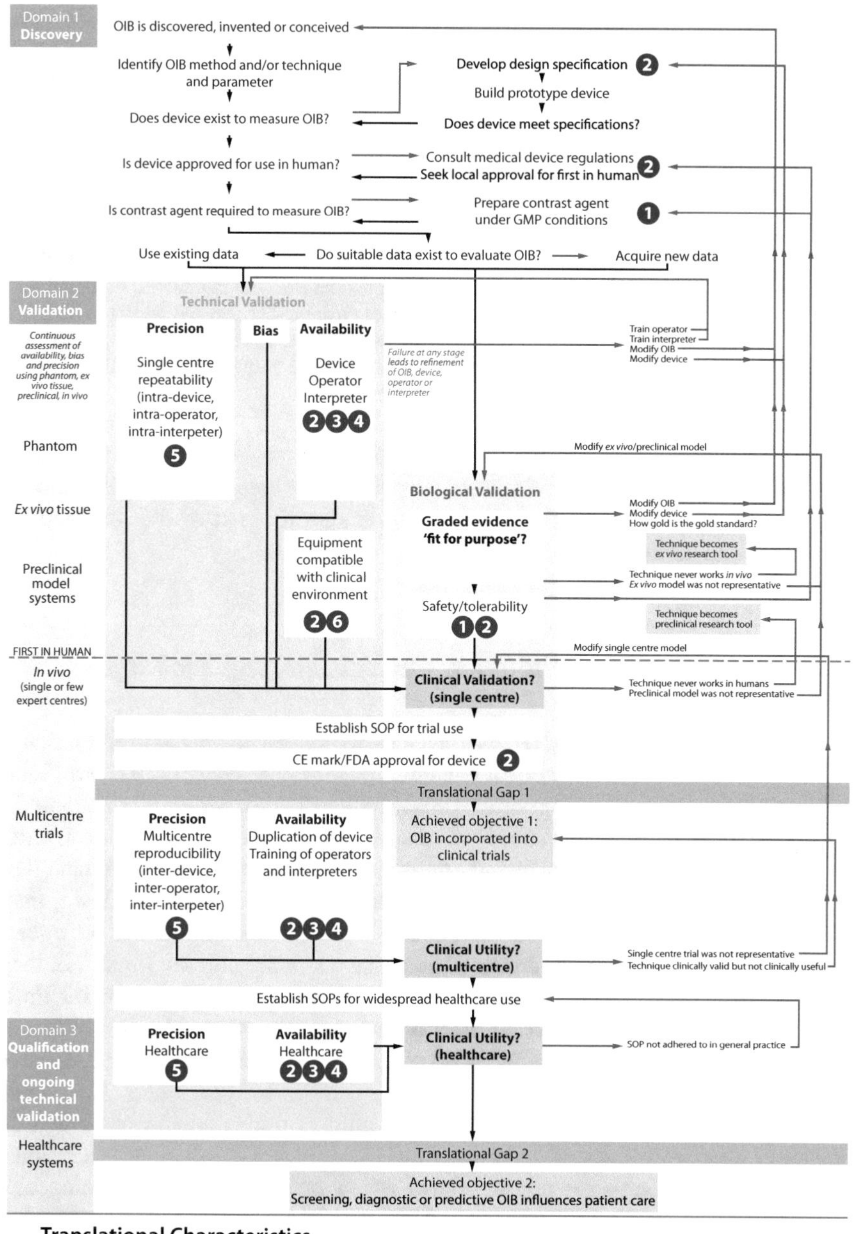

Translational Characteristics

1. Contrast Mechanism
2. Instrumentation
3. Operator Expertise
4. Image Reader Expertise
5. Repeatability and Reproducibility
6. Co-registration

◀**Fig. 1.3** Optical imaging biomarker roadmap. This proposed optical imaging biomarker roadmap differs from the imaging biomarker roadmap [4] by considering the intrinsic coupling of clinical translation to the development of the optical imaging device. Red feedback loops decelerate clinical translation. Purple circles represent points at which careful advance consideration of translational characteristics 1–6 can accelerate translation (Sect. 1.3.2). Translational characteristics are: 1. Contrast Mechanism. 2. Instrumentation. 3. Operator Expertise. 4. Image Reader Expertise. 5. Repeatability and Reproducibility 6. Co-registration. In the optical imaging biomarker roadmap, the technical validation of the technique is more prominent than in the imaging biomarker roadmap, since each new optical imaging technique requires the development of a partially or entirely new optical device. Technical and biological validation occur in parallel and are closely linked due to the interplay between device, operator, interpreter and optical imaging biomarker. Essential technical validation occurs throughout the roadmap to ensure availability and precision in all settings from phantoms to healthcare. Cost-effectiveness is omitted here but impacts on the roadmap at every stage, owing to the equipment and personnel costs of performing imaging studies [4]. SOP: standard operating procedure, OIB: optical imaging biomarker

will be referred to throughout this thesis, emphasizing where their consideration has influenced the methodologies adopted, and outlining the part of the roadmap our work intends to address.

1.3.1 Barriers to Translation of Optical Imaging Techniques

Translational Barrier 1. Lack of defined safe exposure limits for diagnostics
Visible and near-infrared light is not ionising, (H ionisation occurs at ~120 nm), but can lead to thermal damage, photosensitisation and photoallergic reactions under certain conditions [46]. Molecularly damaging radiation extends further towards the visible spectrum [47], so safe and effective illumination levels, those that allow accurate measurement of a given biomarker, at a safe optical radiation level, must be established before in vivo trials take place. The published guidelines for optical radiation exposure limits only concern the retina and skin, and are limited in their application to workers rather than patients, which presents a challenge for those using incoherent light and internal diagnostic imaging devices [46, 48]. There is an unmet need for experimentally validated exposure limits for other tissues so that safe patient exposures for optical imaging can be clearly defined. This would help to address frequently raised concerns of clinical trials ethical review boards. Furthermore, the availability of optical radiation protection advisors with training in clinical optical imaging to advise ethical review boards would be advantageous, to increase understanding of the capabilities, strengths and weaknesses of optical imaging systems and their biomarkers.
Translational Barrier 2. Lack of standardised quality assurance
The uncertainty regarding optical exposure limits is illustrative of a more general lack of standardisation in optical imaging. Defining the performance of a device at the point of manufacture and throughout lifetime is crucial to enable comparison of

different devices measuring the same source of optical contrast and debug device-specific problems.
Whilst performance standards (documents that suggest relevant performance characteristics and test methods) exist for the established non-optical imaging methods, such as the National Electrical Manufacturers Association (NEMA) imaging standards for positron emission tomography (PET) [49], optical imaging is currently without published standards [50]. As a result, optical imaging contrast agents and devices must often be approved in tandem. This is in contrast to PET where contrast agents are usually approved for use in a range of PET devices, and vice versa, due to sufficient evidence supporting the similarity across PET devices [50]. Unfortunately, the consequence for optical imaging is that healthcare providers must purchase specific devices when working with specific optical imaging agents, resulting in prohibitive costs and blocking translation of new agents.
In the US and Europe, medical devices may be approved based on similarity to a predicate device, gaining 510(k) clearance in the US and CE marking in the EU [51, 52]. But without proper standardisation, similarity to a predicate may be difficult to define, leaving performance unclear. Even worse, the predicate could predate the introduction of regulatory standards. With proper imaging standards, device performance could be properly compared, allowing inferiority to be spotted at an earlier stage in the translational roadmap. Additionally, if device performance standards were established it would help separate device approval from agent approval, in cases where exogenous agents are being imaged.
Translational Barrier 3. Lack of accurate, validated, clinical gold standards
New techniques are often compared to clinical gold standards to measure clinical performance, but these are a common source of problems limiting translation. Most commonly, gold standard diagnosis is determined by assessment of stained tissue sections by a pathologist. To achieve the most accurate gold standard for translation of optical imaging biomarkers, consensus of several independent pathologists is needed, as they are not always in agreement [53]. In the best-case scenario, only unanimous decisions would be accepted, though this is difficult to achieve in practice as it significantly decreases the number of samples that can be incorporated into any calculation. In the future, machine learning algorithms trained on huge datasets may be able to provide an objective diagnosis, but this is still some way from being realised.
Another confounding factor is the transition from a biopsy read to in vivo imaging, which requires an appreciation of the nuances of a field of tissue being a mix of pathology as well as normal anatomy. Each optical imaging pixel could include contributions from one or more tissue states, whereas a histopathology analysis will typically report only the highest grade present. In addition, high congruence in spatial alignment between the in vivo and ex vivo coordinates is required. Once the gold standard is validated, careful thought must be applied to how these findings will be accurately co-registered with optical images, so as not to introduce further artefacts into the comparison between the novel technique and the gold standard. The different scales at which in vivo optical image data is recorded and histopathological analysis is performed can make this particularly challenging [54]. Throughout this thesis,

where novel techniques are compared to gold standards, methods were carefully designed to simplify coregistration.

Translational Barrier 4. Lack of validated, representative ex vivo models

To avoid disappointing in vivo results later in the roadmap, it is also crucial to ensure that the ex vivo model is as representative as possible of the in vivo situation. When tissue is excised from the body, several OIBs change irreversibly. For example, the lack of active blood flow changes the spectrum of tissue by reducing blood oxygenation to 0% and consequently altering the haemoglobin absorption spectrum. Furthermore, tissue autofluorescence can be modified upon exposure to ambient light and tissue structure may be distorted by surgical trauma, or by positioning the tissue on a rigid surface. Ultimately, tissue will degrade unless fixed in formaldehyde or frozen, which further alter properties [55]. The leap between data acquired ex vivo and in vivo imaging is therefore large and data acquired from ex vivo tissues may contain insurmountable artefacts if the tissues are not properly handled.

These challenges with using ex vivo tissue for validation raise the need for new model systems, which may arise through improved tissue mimicking phantoms, or from bioengineering of artificial tissues. Whichever route is chosen, to avoid disappointing in vivo results later in the roadmap, it is crucial to ensure that the ex vivo model is as representative as possible of the in vivo situation. For example, modalities that are sensitive to blood oxygenation should never use excised tissue for validation unless it is perfused, while modalities sensitive to preparation artefacts should avoid chemically fixed or frozen tissue [55] and dyes that incorporate labelled human antibodies should use models that try to replicate human biology, such as human derived tumour xenografts, rather than animal tumour models where the binding affinity may be different.

Translational Barrier 5. Difficulty in conducting representative single centre trials

Similarly, early stage single centre trials should replicate the common clinical environment as far as possible. Firstly, representative populations should be chosen to reduce spectrum effects [56]. For example, due to the ethical considerations of taking biopsies from healthy tissue, many skin imaging trials have been carried out in enriched populations with high disease incidence, which has prompted investigators to endorse the need for a mix of lesions representative of the target population to be used for future testing of new approaches [57] and encouraging regulatory bodies to disregard earlier results. Secondly, standard operating procedures (SOPs) should be determined by a single centre and adhered to in multi-centre trials in order to prevent bias. For example, inspecting images for minutes when the eventual SOP in a clinical setting would require inspection of videos in real time gives misleading performance evaluations [58–61]. Thirdly, the expertise of the operator and interpreter should be representative of that realistically achieved in routine care. Endoscopic trimodal imaging (Sect. 2.2.2) provides a cautionary example where promising results in a specialist tertiary referral centre [62] were not reproduced in a community practice setting [63], which can be due to a combination of spectrum effects, different SOPs and different expertise.

1.3.2 *Translational Characteristics of Optical Imaging Techniques*

In constructing the OIB Roadmap, we identified several 'translational characteristics' (TCs) that seem to be common among those techniques that have been widely adopted. These characteristics do not define technical performance (such as resolution or signal to noise ratio) or clinical performance (such as negative predictive value or specificity) but rather characterise the amenability of a technique to clinical translation. The OIB Roadmap highlights the key points at which consideration of these translational characteristics can be exploited to accelerate translation (Fig. 1.3). The translational characteristics are detailed below together with suggestions for how to best achieve translation.

Translational Characteristic 1. Contrast Mechanism. Exploiting endogenous contrast (imaging without application of dyes) to derive the OIB is favourable in terms of the clinical pathway. Still, exogenous contrast agents can be beneficial for improving the contrast of cancer compared to healthy tissue, both as non-specific stains [64] and targeted molecular imaging agents [21], but they require synthesis at good manufacturing practice (GMP) standards, associated toxicology studies, and in addition to the need for specific instrumentation, also add procedure time and cost [20, 65]. Several aspects of contrast agent chemistry increase the likelihood of their successful clinical translation: having a validated target (structural or molecular) increases confidence that results will be reproducible; topically rather than intravenously administered agents limit the exposure to the tissue of interest and speed up procedures; and agents with long term stability once formulated are favourable for storage and distribution.

Translational Characteristic 2. Instrumentation. A regulatory body such as the FDA must approve new devices [50, 51, 66, 67]. For clinical implementation, compact, robust and transportable optical imaging devices devoid of complex delicate optics are highly desirable. Devices that are compatible with existing systems, or include current standard of care methods for reference, are more likely to be translated, perhaps initially as an adjunct to an existing technique, which also facilitates head-to-head trials.

Translational Characteristic 3. Operator Expertise. The potential operators of a new optical imaging technique range from the untutored public through professionals working in primary care to highly specialised individuals working in a specialist care centre. A new approach must either: give similar clinical results to an existing approach, but with less expertise required; reduce the required expertise sufficiently to allow translation of the approach from expert to generalist setting, reducing the burden on specialist care centres and reducing the cost of running a high-volume imaging suite [68]; or improve the existing standard of care sufficiently to justify an increased level of specialist knowledge required.

Translational Characteristic 4. Image Reader Expertise. Converting the raw image data into a relevant OIB is a crucial step, whereby image interpretation criteria must be established. Such criteria need to deliver high sensitivity, specificity, inter-observer agreement and short learning curves. Criteria can include a binary decision, a library-based classification, the presence of specific image patterns [69, 70], or a change in signal intensity relative to a defined threshold. Establishing OIBs is a time-consuming task, often requiring international consensus across multiple centres.

For more complex biomarkers, expert image readers need to be trained, which further adds to the cost and time for translation, as well as making the biomarker difficult to standardise across centres. Providing simple feedback of the biomarker, or familiar images that avoid the need for retraining, can enable non-experts to make diagnoses and smooth the translational pathway.

Partially or fully automated analysis provides an attractive means of reducing or removing the burden on expert image readers. It has the potential to be objective, standardised and low-cost, but has yet to mature to stage where it is fully capable of operating in real time with sufficient performance to replace the human image reader [71, 72]. For many optical imaging techniques, data reduction is essential because the dimensionality of the data is beyond interpretation by the human image reader.

Translational Characteristic 5. Repeatability and Reproducibility. Both repeatability (test–retest) and reproducibility across devices, operators and interpreters must be assessed to evaluate the achievable precision for measurement of the OIB. Though inter-observer agreement (encompassing operators and interpreters) is often assessed, the intra-observer, intra-device and inter-device variability are often overlooked, which makes comparative evaluation of novel OIBs challenging. To maximise the opportunity for translation to the intended setting, studies should be designed to enable comparison of results obtained across multiple centres.

Translational Characteristic 6. Co-registration. A key facet of imaging is the ability to provide spatially resolved information which can be co-registered with the anatomy of the patient to allow guidance for intervention. There are several methods of achieving this with increasing levels of complexity that could hinder translation: co-registration with an existing technique that is compatible with surgical treatment [73]; application of fiducial markers, such as laser cautery marks, to highlight target areas [74]; or projection of the image data onto the patient or into the surgeon's field of view using augmented reality technology [75, 76].

1.4 Outlook

Imaging is the only medical tool currently capable of non-invasively capturing detailed, immediate and spatially resolved biochemical information in vivo and thus delineating disease such that non-invasive curative resection or treatment can take place. Despite great promise, few optical imaging techniques have been translated to routine clinical use. The translational characteristics we identified complement the OIB Roadmap, which we hope will improve the chances of novel optical imaging techniques being translated to widespread clinical implementation. With such a vast array of complex tissue-light interactions, and an equally diverse arsenal of developing optical devices, some are sure to change the current standard of care for cancer patients, leading to improved outcomes in the future.

This thesis describes recent efforts to advance three optical endoscopic imaging techniques towards clinical translation. It is split into two parts: Chaps. 2–5 describe work on two novel flexible endoscopic imaging techniques for early detection of cancer in the oesophagus; Chap. 6 describes work on a novel rigid endoscopic imaging technique for delineation of cancer and healthy tissue in surgery to remove pituitary tumours. For each of these techniques, the following aims were established.

- Guided by the translational characteristics, to develop a novel optical imaging technique capable of swift clinical translation, including development of the device, contrast mechanism and image analysis methods.
- To translate the technique through the beginning of Domain 2 of the OIB Roadmap, including technical characterisation, to ensure the device meets design specifications, and technical validation, to ensure precision, and biological validation, to ensure the device is able to distinguish between the tissues of interest in a range of representative ex vivo samples.
- To ultimately conduct the first-in-human trials of the technique, including design and approval of a clinical pilot study and local safety approval of the device and contrast mechanism for use in humans.

This thesis concludes with an outlook for the future, outlining the steps required to improve each technique and evaluating their potential to advance further towards clinical translation and to ultimately improve the standard of care in their respective indications.

In the following chapters, reference to the translational characteristics and translational barriers will frequently be made to highlight where their consideration has influenced the successes and failures of optical imaging techniques, as well as to underline where they have influenced the decisions in designing the techniques and experimental methodologies described in this thesis. Reference to the OIB Roadmap shall also be made to emphasise where each stage of the work sits on the path to translation and highlight the experiments required to address the next steps in the roadmap.

References

1. Wax A, Terry NG, Dellon ES, Shaheen NJ (2011) Angle-resolved low coherence interferometry for detection of dysplasia in Barrett's esophagus. Gastroenterology 141:443–447
2. Bigio IJ, Fantini S (2016) Quantitative biomedical optics. Cambridge University Press
3. Imamoto Y, Shichida Y (2014) Cone visual pigments. Biochim Biophys Acta Bioenerg 1837:664–673
4. O'Connor JPB et al (2017) Imaging biomarker roadmap for cancer studies. Nat Rev Clin Oncol 14:169–186
5. Hanahan D, Weinberg RA (2011) Hallmarks of cancer: the next generation. Cell 144:646–674

6. Bhat YM et al (2014) High-definition and high-magnification endoscopes. Gastrointest Endosc 80:919–927
7. Rameshshanker R, Wilson A (2016) Electronic imaging in colonoscopy: clinical applications and future prospects. Curr Treat Options Gastroenterol 14:140–151
8. Manfredi MA et al (2015) Electronic chromoendoscopy. Gastrointest Endosc 81:249–261
9. Thosani N et al (2016) ASGE technology committee systematic review and meta-analysis assessing the ASGE preservation and incorporation of valuable endoscopic innovations thresholds for adopting real-time imaging–assisted endoscopic targeted biopsy during endoscopic surveillance. Gastrointest Endosc 83:684–698
10. Lu G, Fei B (2014) Medical hyperspectral imaging: a review. J Biomed Opt 19:10901
11. Jang J (2015) The past, present, and future of image-enhanced endoscopy. 466–475
12. Olympus-Technologies NBI | Medical Systems. Available at https://www.olympus-europa.com/medical/en/medical_systems/technologies/narrow_band_imaging__nbi_1/technologies_nbi.jsp. Accessed on 26th Apr 2017
13. FICE Dual Mode | Fujifilm Europe. Available at https://www.fujifilm.eu/eu/products/medical-systems/endoscopy/technology/fice-dual-mode. Accessed on 26th Apr 2017
14. Advanced Imaging | PENTAX Medical (EMEA). Available at https://www.pentaxmedical.com/pentax/en/95/1/WavSTAT4-Optical-Biopsy-System. Accessed on 26th Apr 2017
15. MedX Health—About SIMSYS-MoleMate™. Available at http://medxhealth.com/Our-Products/SIAscopytrade;/overview.aspx. Accessed on 26th Apr 2017
16. MelaFind. Available at http://www.melafind.com/melafind/. Accessed on 26th Apr 2017
17. Spectral Molecular Imaging—Products. Available at http://www.opmol.com/products.html. Accessed on 26th Apr 2017
18. Alali S, Vitkin A (2015) Polarized light imaging in biomedicine: emerging Mueller matrix methodologies for bulk tissue assessment. J Biomed Opt 20:061104
19. [DERMA MEDICAL SYSTEMS] English. Available at https://www.dermamedicalsystems.com/index.php?menu_id=117. Accessed on 2nd May 2017
20. Sturm MB, Wang TD (2015) Emerging optical methods for surveillance of Barrett's oesophagus. Gut 64:1816–1823
21. James ML, Gambhir SS (2012) A molecular imaging primer: modalities, imaging agents, and applications. Physiol Rev 92:897–965
22. Ulrich M et al (2016) Dynamic optical coherence tomography in dermatology. Dermatology 232:298–311
23. Cellvizio: Our Flagship Product | Mauna Kea Technologies. Available at http://www.maunakeatech.com/en/hospital-administrators/cellvizio-solution. Accessed on 26th Apr 2017
24. ViewnVivo Home—ViewnVivo—Must-see, miniaturised in vivo microscopy. Available at http://viewnvivo.com/. Accessed on 26th Apr 2017
25. Caliber ID. Clinical applications. Available at http://www.caliberid.com/clinical.html. Accessed 2nd May 2017
26. Song LMWK et al (2011) Autofluorescence imaging. Gastrointest Endosc 73:647–650
27. Hanson KM, Bardeen CJ (2009) Application of nonlinear optical microscopy for imaging skin. Photochem Photobiol 85:33–44
28. Fu L, Gu M (2007) Fibre-optic nonlinear optical microscopy and endoscopy. J Microsc 226:195–206
29. Thomas G, Van Voskuilen J, Gerritsen HC, Sterenborg HJCM (2014) Advances and challenges in label-free nonlinear optical imaging using two-photon excitation fluorescence and second harmonic generation for cancer research. J Photochem Photobiol, B 141:128–138
30. Jenlab: DermaInspect. Available at http://www.jenlab.de/DermaInspect.29.0.html. Accessed on 5th May 2017
31. Trindade AJ, Smith MS, Pleskow DK (2016) The new kid on the block for advanced imaging in Barrett's esophagus: a review of volumetric laser endomicroscopy. Ther Adv Gastroenterol 9:408–416
32. NinePoint Medical. Available at http://www.ninepointmedical.com/#NvisionVLE. Accessed on 26th Apr 2017

33. Verisante Technology, Inc. Available at http://www.verisante.com/products/core/. Accessed on 26th Apr 2017
34. Boone MALM, Norrenberg S, Jemec GBE, Del Marmol V (2014) High-definition optical coherence tomography imaging of melanocytic lesions: a pilot study. Arch Dermatol Res 306:11–26
35. Product—Vivosight. Available at https://vivosight.com/about-us/product/. Accessed 28th Apr 2017
36. Kallaway C et al (2013) Advances in the clinical application of Raman spectroscopy for cancer diagnostics. Photodiagn Photodyn Ther 10:207–219
37. Pence I, Mahadevan-Jansen A (2016) Clinical instrumentation and applications of Raman spectroscopy. Chem Soc Rev 45:1958–1979
38. Wang W, Zhao J, Short M, Zeng H (2015) Real-time in vivo cancer diagnosis using raman spectroscopy. J Biophotonics 8:527–545
39. Tu Q, Chang C (2012) Diagnostic applications of Raman spectroscopy. Nanomed: Nanotechnol Biol Med 8:545–558
40. Verisante Aura^{TM}. Available at http://www.verisante.com/aura/medical_professional/. Accessed on 26th Apr 2017
41. Benes Z, Antos Z (2009) Optical biopsy system distinguishing between hyperplastic and adenomatous polyps in the colon during colonoscopy. Anticancer Res 29:4737–4739
42. Boerwinkel DF et al (2014) Fluorescence spectroscopy incorporated in an optical biopsy system for the detection of early neoplasia in Barrett's esophagus. Dis Esophagus : Off J Int Soc Dis Esophagus/I.S.D.E
43. Grosenick D, Rinneberg H, Cubeddu R, Taroni P (2016) Review of optical breast imaging and spectroscopy. J Biomed Opt 21:091311
44. Yu G (2012) Near-infrared diffuse correlation spectroscopy in cancer diagnosis and therapy monitoring. J Biomed Opt 17:010901
45. Sharma P et al (2015) White paper AGA: advanced imaging in Barrett's esophagus. Clin Gastroenterol Hepatol 13:2209–2218
46. ICNIRP (2013) ICNIRP guidelines on limits of exposure to incoherent visible and infrared radiation. Health Phys 71:804–819
47. Brookner CK, Agrawal A, Trujillo EV, Mitchell MF, Richards-Kortum RR (1997) Safety analysis: relative risks of ultraviolet exposure from fluorescence spectroscopy and colposcopy are comparable. Photochem Photobiol 65:1020–1025
48. Directive 2006/25/EC of the European Parliament and of the Council (2006)
49. NEMA (2013) Performance measurements of positron emission tomographs (PETs). Available at https://www.nema.org/Standards/Pages/Performance-Measurements-of-Positron-Emission-Tomographs.aspx
50. Tummers WS et al (2017) Regulatory aspects of optical methods and exogenous targets for cancer detection. Can Res 77:2197–2206
51. Van Norman GA (2016) Drugs and devices: comparison of European and U.S. approval processes. JACC: Basic Transl Sci 1:399–412
52. Van Norman GA (2016) Drugs, devices, and the FDA: Part 1. JACC: Basic Transl Sci 1:170–179
53. Downs-Kelly E et al (2008) Poor interobserver agreement in the distinction of high-grade dysplasia and adenocarcinoma in pretreatment Barrett's esophagus biopsies. Am J Gastroenterol 103:2333–40; quiz 2341
54. O'Connor JPB et al (2015) Imaging intratumor heterogeneity: Role in therapy response, resistance, and clinical outcome. Clin Cancer Res 21:249–257
55. Shim MG, Wilson BC (1996) The effects of ex vivo handling procedures on the near-infrared Raman spectra of normal mammalian tissues. Photochem Photobiol 63:662–671
56. Usher-Smith JA, Sharp SJ, Griffin SJ (2016) The spectrum effect in tests for risk prediction, screening, and diagnosis. BMJ. i3139
57. Monheit G et al (2011) The performance of MelaFind: a prospective multicenter study. Arch Dermatol 147:188–194

58. Silva FB et al (2011) Endoscopic assessment and grading of Barrett's esophagus using magnification endoscopy and narrow-band imaging: Accuracy and interobserver agreement of different classification systems (with videos). Gastrointest Endosc 73:7–14
59. Kara MA, Ennahachi M, Fockens, P, ten Kate FJW, Bergman JJGHM (2006) Detection and classification of the mucosal and vascular patterns (mucosal morphology) in Barrett's esophagus by using narrow band imaging. Gastrointest Endosc 64, 155–66
60. Singh R et al (2008) Narrow-band imaging with magnification in Barrett's esophagus: validation of a simplified grading system of mucosal morphology patterns against histology. Endoscopy 40:457–463
61. Sharma P et al (2006) The utility of a novel narrow band imaging endoscopy system in patients with Barrett's esophagus. Gastrointest Endosc 64:167–175
62. Curvers WL et al (2010) Endoscopic tri-modal imaging is more effective than standard endoscopy in identifying early-stage neoplasia in Barrett's esophagus. Gastroenterology 139:1106–1114
63. Curvers WL et al (2011) Endoscopic trimodal imaging versus standard video endoscopy for detection of early Barrett's neoplasia: a multicenter, randomized, crossover study in general practice. Gastrointest Endosc 73:195–203
64. Trivedi PJ, Braden B (2013) Indications, stains and techniques in chromoendoscopy. QJM: Mon J Assoc Phys 106:117–131
65. Sevick-Muraca EM et al (2013) Advancing the translation of optical imaging agents for clinical imaging. Biomed Opt Express 4:160–170
66. Overview of device regulation. Available at https://www.fda.gov/MedicalDevices/DeviceRegulationandGuidance/Overview/default.htm. Accessed on 27th Nov 2017
67. Medical devices—European Commission. Available at https://ec.europa.eu/growth/sectors/medical-devices_en. Accessed on 27th Nov 2017
68. Gora MJ et al (2013) Tethered capsule endomicroscopy enables less invasive imaging of gastrointestinal tract microstructure. Nat Med 19:238–240
69. Sharma P et al (2016) Development and validation of a classification system to identify high-grade dysplasia and esophageal adenocarcinoma in Barrett's esophagus using narrow band imaging. Gastroenterology 150:591–598
70. Boerwinkel DF et al (2014) Third-generation autofluorescence endoscopy for the detection of early neoplasia in Barrett's esophagus: a pilot study. Dis Esophagus: Off J Int Soc Dis Esophagus/I.S.D.E 27:276–84
71. de Bruijne M (2016) Machine learning approaches in medical image analysis: from detection to diagnosis. Med Image Anal 33:94–97
72. Suzuki K (2017) Overview of deep learning in medical imaging. Radiol Phys Technol 10:1–17
73. Garcia-Allende PB et al (2013) Towards clinically translatable NIR fluorescence molecular guidance for colonoscopy. Biomed Opt Express 5:78–92
74. Suter MJ et al (2014) Esophageal-guided biopsy with volumetric laser endomicroscopy and laser cautery marking: a pilot clinical study. Gastrointest Endosc 79:886–896
75. Pelargos PE et al (2016) Utilizing virtual and augmented reality for educational and clinical enhancements in neurosurgery. J Clin Neurosci 35:1–4
76. Nicolau S, Soler L, Mutter D, Marescaux J (2011) Augmented reality in laparoscopic surgical oncology. Surg Oncol 20:189–201

Chapter 2
Flexible Endoscopy: Early Detection of Dysplasia in Barrett's Oesophagus

Early detection of cancer confers more than a three-fold increase in ten-year-survival rate [1], but screening is challenging: it must achieve low false negative rates to avoid missing potentially deadly disease, whilst retaining low false positive rates to prevent over-treatment; it must be low-cost such that it can be widely implemented and crucially, if localised treatment is to be possible, for example by ablation or resection, the screening technique must provide spatially resolved information. Optical imaging has the potential to achieve these goals by capturing spatially resolved biochemical information based on the contrast mechanisms discussed in Chap. 1 and summarised in Fig. 1.1.

Though optical imaging is relatively low cost in comparison to radiological imaging, widespread screening regimes can be prohibitively expensive. Fortunately, some cancers are associated with a known precursor disease, enabling us to more affordably perform surveillance on a smaller enriched population, detect the early stages of cancer, non-invasively resect or treat this, and thus improve survival rates. One such disease is Barrett's oesophagus. This chapter describes Barrett's oesophagus, the current standard of care surveillance, and the current state of the art in advanced endoscopic surveillance techniques.

2.1 Barrett's Oesophagus

Barrett's oesophagus is an acquired condition in which columnar epithelium replaces the stratified squamous epithelium of the lining of the distal oesophagus, the pipe which connects the throat to the stomach. Crucially for early detection attempts, Barrett's oesophagus predisposes patients to the development of oesophageal adenocarcinoma (OAC) [2]. The progression to cancer occurs through an intermediate stage known as dysplasia, which can be of low-grade (LGD) or high-grade (HGD). The elevated cancer risk in Barrett's oesophagus is variably reported (Fig. 2.1) but is somewhere around 0.1–0.3%/year in non-dysplastic Barrett's oesophagus [3–5], 9%/year in the presence of LGD [6] and around 4 times higher in patients

D. J. Waterhouse, *Novel Optical Endoscopes for Early Cancer Diagnosis and Therapy*, Springer Theses, https://doi.org/10.1007/978-3-030-21481-4_2

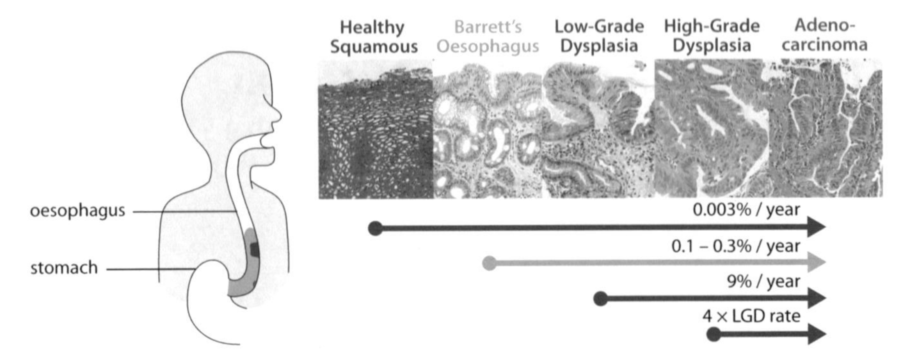

Fig. 2.1 Progression of Barrett's oesophagus. H&E stained histopathology images illustrate the histopathological cell features that define Barrett's oesophagus, dysplasia and cancer in the distal oesophagus. In Barrett's oesophagus, healthy squamous epithelium is replaced by columnar epithelium. Low-grade dysplasia shows architectural and cytological abnormalities associated with neoplasia, such as crowded crypts and enlarged nuclei. In high-grade dysplasia these are more prominent. In adenocarcinoma, the neoplastic cells penetrate through the basement membrane into the lamina propria. Reproduced from [22]. Progression figures for Barrett's oesophagus are difficult to accurately determine, but estimates are shown [3–7]

harbouring HGD, compared to patients with LGD [7]. As the 5-year survival rate for oesophageal cancer is just 15%, but improves to 80% when the cancer is identified at an early-stage [8, 9], major advisory bodies recommend that patients with Barrett's oesophagus undergo routine endoscopic surveillance for signs of dysplasia or early carcinoma [10–14] such that they can be treated non-invasively with curative intent.

Indeed, data from some retrospective studies indicate that surveillance correlates with improved survival [15–17], although evidence from a case-controlled study [18] did not confirm this and data from randomised controlled trials is lacking. Still, patients currently undergo surveillance endoscopy every 3–5 years.

The current standard of care (SOC) for endoscopic surveillance uses high-definition white light endoscopy (HD-WLE) to identify suspicious lesions based on their appearance. Lesions are often difficult to spot, as they are heterogeneous in their shape and size, patchy in their distribution and, especially when flat, show subtle contrast on HD-WLE. Suspicious lesions are biopsied, stained and sectioned to reveal the cell morphology, which is analysed by a pathologist to reach a diagnosis (Fig. 2.2).

To mitigate the risk of missing subtle, flat lesions, the SOC protocol also includes taking random biopsies. These are taken at four-quadrant positions circumferentially every 2 cm along the region of Barrett's oesophagus, a process known as the Seattle protocol, in the hope that this frequent sampling will identify any lesions missed by HD-WLE [10, 19]. The resulting sensitivity is 40–64% with specificity of 98–100% [20], but the procedure is costly, time-consuming and prone to sampling error [14].

The potential to improve clinical outcomes by increasing contrast for dysplasia with advanced optical techniques has driven a great deal of research in this area. With demand for endoscopy predicted to rise substantially over the next decade [21],

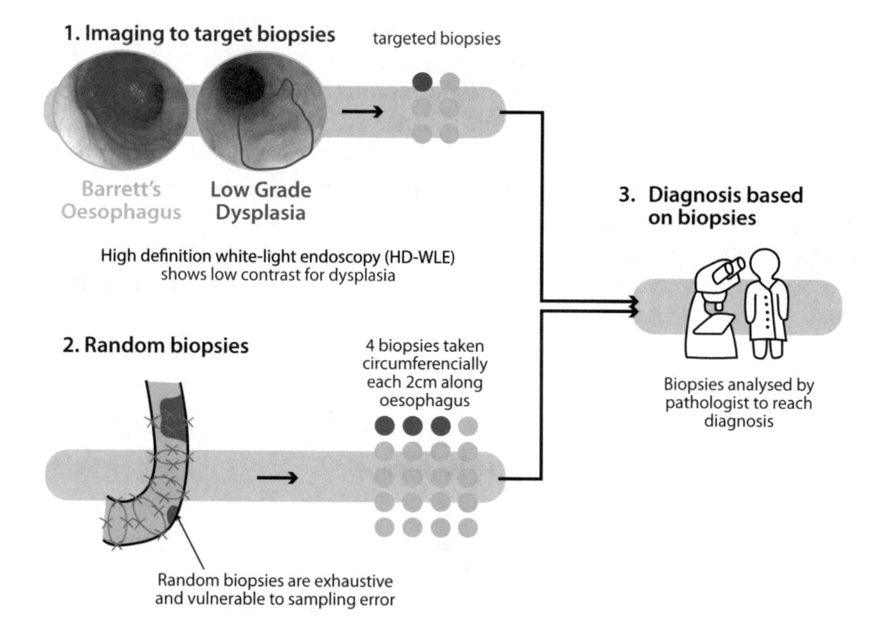

Fig. 2.2 Current standard of care for endoscopic surveillance of patients with Barrett's oesophagus. Biopsies are targeted by inspection of the oesophagus using high definition white-light endoscopy, but this is limited by poor contrast for dysplasia. Random biopsies are taken at four-quadrant positions circumferentially every 2 cm along the oesophagus, a process known as the Seattle protocol, in the hope that this frequent sampling will identify any lesions missed by HD-WLE, but these are time consuming, costly and prone to sampling error. Final diagnosis is made from histopathology of the biopsies. These limitations have motivated research into advanced optical techniques for endoscopic surveillance

the unmet clinical need for optical methods with improved diagnostic yield and/or lower cost per procedure is particularly acute. As discussed in Sect. 1.1, harnessing the complex interactions between light and tissue in a clinically translatable optical imaging technique has tremendous potential for non-invasive detection and characterisation of disease tissue, in this case, detection of dysplasia.

2.2 Improving the Standard of Care

2.2.1 Categories of Advanced Optical Endoscopic Techniques

Advanced optical endoscopic techniques are often categorised as 'red-flag', 'optical biopsy' or 'hybrid' (Fig. 2.3). Red-flag techniques provide wide-field images and, if they provide sufficient contrast for dysplasia, can replace both HD-WLE and random four-quadrant biopsies through improved targeting of biopsies (Fig. 2.3b). A recent study estimated that using a targeted biopsy protocol without random biopsies could reduce per-patient biopsy costs from ~£1000 to ~£30 [23]. Conversely, optical biopsy techniques measure a small area of tissue with the goal of providing in vivo, real-time diagnosis (Fig. 2.3c). In the short-term, this could reduce the number of biopsies by pre-screening prospective biopsy locations. In the long-term, optical biopsy could replace physical biopsy, enabling immediate diagnosis during surveillance, allowing intervention to occur directly within the same procedure. Hybrid techniques, as the name suggests, combine red flag and optical biopsy capabilities to identify and diagnose disease in vivo.

2.2.2 Advanced Optical Endoscopic Techniques in Clinical Practice for Barrett's Surveillance

Several advanced optical techniques are already in clinical use for endoscopic Barrett's surveillance in some centres. Still, endoscopic practice varies significantly across countries and within the same country, and use of these advanced optical techniques is often restricted to tertiary referral centres delivering endoscopic treatment to a high volume of dysplastic patients. A recent meta-analysis by the American Society for Gastrointestinal Endoscopy (ASGE) suggest that to be recommended for targeting biopsy, a new technique should achieve at least 90% sensitivity, 80% specificity and 98% negative predictive value [24]. Table 2.1 shows the recommendation status of techniques currently used in clinic, as well as their key advantages and disadvantages. Figure 2.4 shows example images from each of these techniques. As they have been extensively reviewed elsewhere [25, 26], only a brief summary of each technique is presented here.

2.2.2.1 Chromoendoscopy (Red Flag)

Chromoendoscopy enhances contrast through the use of topically applied dyes. Acetic acid eliminates the superficial mucosal layer and then causes changes to (acetylation of) cellular proteins, resulting in whitening that highlights surface patterns. In the case of neoplastic Barrett's, this is rapidly followed by focal erythema (redness) caused by vascular congestion in capillaries, which is revealed as focal redness as loss of acetowhitening occurs [23]. These reactions are used to guide targeted biopsies and increase the yield of dysplasia, meeting the ASGE performance thresholds [24, 40]. Methylene blue chromoendoscopy has also been extensively investigated, but there are concerns regarding possible carcinogenic effects of the dye [27]. Meta-analyses have found it to be inferior to WLE [41] and acetic acid chromoendoscopy [24]. It is therefore likely that acetic acid will become the standard conventional chromoendoscopic method for Barrett's surveillance.

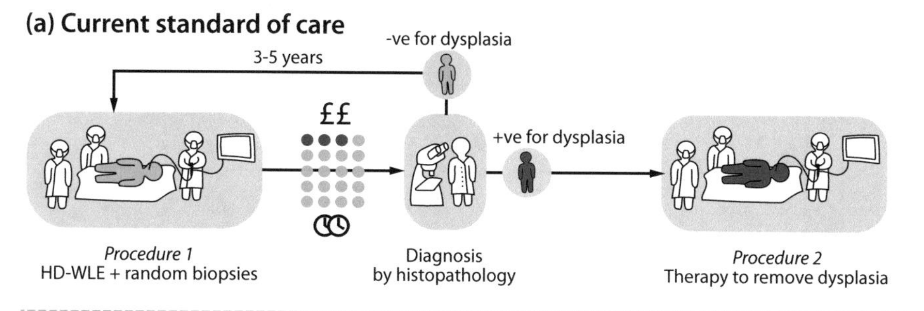

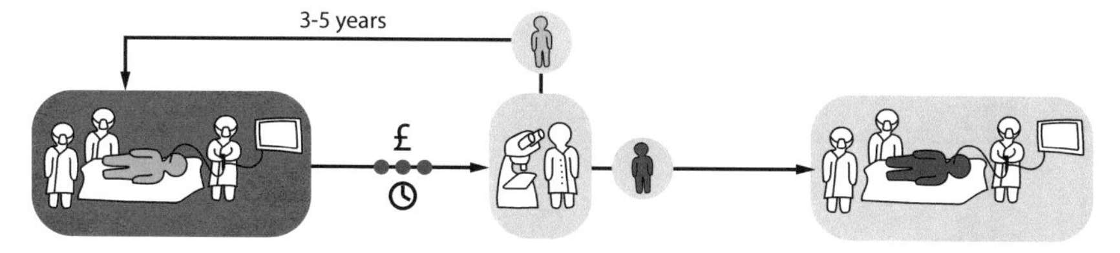

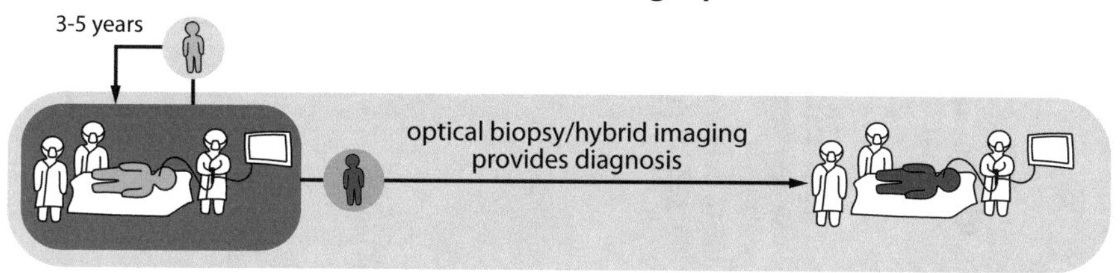

Fig. 2.3 'Red flag', 'optical biopsy' and 'hybrid' techniques in relation to the standard of care. **a** The current standard of care results in a delay between surveillance endoscopy, diagnosis and treatment, and is costly due to the exhaustive biopsies taken in an attempt to overcome the limitations of red flag imaging with HD-WLE. **b** Improved 'red flag' imaging would survey the oesophagus, identifying suspicious regions with greater sensitivity and specificity than HD-WLE and random biopsies, thus reducing the number of biopsies. **c** 'Optical biopsy' might one day replace physical biopsy providing real time diagnosis in vivo and thus removing the need for physical biopsies and allowing surveillance and treatment to occur in a single procedure. Optical biopsy techniques must be paired with a red flag technique to allow wide field surveillance. Alternatively, 'hybrid' techniques perform the role of both red flag and optical biopsy by scanning the entire oesophagus and providing diagnostic information in vivo

Table 2.1 Advanced optical techniques in clinical use for endoscopic surveillance of Barrett's oesophagus

Technique	Advantages	Disadvantages	Sensitivity	Specificity	Recommendation				
					BSG	ASGE[a]	ACG	AGA	ESGE
HD-WLE + 4QB and histopathology of biopsies	Widely available; well established	Prone to sampling error [14]; exhaustive biopsies are expensive	0.40–0.68 [20]	0.98–1.00 [20]	✔	✔	✔	✔	✔
Chromoendoscopy	Inexpensive; widely available; acetic acid has shown high sensitivity and specificity for detecting dysplasia [24]	Potential toxicology issues [27] (methylene blue); increase in procedure time [24]; low inter-observer agreement; no current procedural terminology for billing and reimbursement [24]; difficulty in achieving uniform application of dye [14]	0.92–0.97 (acetic acid) 0.642 (methylene blue) [24, 28]	0.85–0.96 (acetic acid) 0.96 (methylene blue) [24, 28]	✗	✔ (acetic acid)	✗	✗	✗

(continued)

Table 2.1 (continued)

Technique	Advantages	Disadvantages	Sensitivity	Specificity	Recommendation				
					BSG	ASGE[a]	ACG	AGA	ESGE
Hardware-based virtual chromoen-doscopy e.g. NBI, BLI	Ability to visualise mucosal and vascular patterns; widely available; ease of use; high sensitivity and specificity for detecting dysplasia [24]; reduced number of biopsies [29]	No universal classification criteria until recent BING criteria [30]; low inter-observer agreement; low sensitivity for low-grade dysplasia [31]	0.94 [24]	0.98 [24]	✗	✔	✗	··	✗
Software-based virtual Chro-moendoscopy e.g. FICE, iSCAN	No additional hardware costs	Lack of data	0.83 [32] (high-grade dysplasia with FICE)	Unavailable	✗	✗	✗	··	✗
Autofluorescence imaging (AFI)	Easy to combine with NBI and WLE	Many studies biased by comparison with substandard WLE [33] Limited value in routine surveillance [33]	0.50 [34] (high-grade dysplasia)	0.61 [34] (high-grade dysplasia)	✗	✗	✗	✗	✗

(continued)

Table 2.1 (continued)

Technique	Advantages	Disadvantages	Sensitivity	Specificity	Recommendation				
					BSG	ASGE[a]	ACG	AGA	ESGE
Endoscopic trimodal imaging (ETMI)	Reduced false positive rate relative to AFI alone	Useful in tertiary referral centres [35] but not in community practice [36]	0.81 [24]	0.46 [24]	✗	✗	✗	✗	✗
Probe-based confocal laser endomicroscopy (pCLE)	Probe can be inserted through working channel of standard endoscope; close to in-vivo histology	Often uses exogenous contrast (fluorescein)	0.90 [24]	0.77 [24]	✗	✗	✗	✗	✗
Endoscope-based confocal laser endomicroscopy (eCLE)	Close to in-vivo histology; high sensitivity and specificity for detecting dysplasia [24]	Requires dedicated endoscope (in contrast to pCLE); often uses exogenous contrast (fluorescein)	0.90 [24]	0.93 [24]	✗	✔	✗	✗	✗

Example images for each technique are shown in Fig. 2.4
BSG British Society of Gastroenterology; *ASGE* American Society for Gastrointestinal Endoscopy; *ACG* American College of Gastroenterology; *AGA* American Gastroenterology Association; *ESGE* European Society of Gastrointestinal Endoscopy
·· Cannot be advocated or discouraged at this time. [a]The ASGE systematic review and meta-analysis indicates that these modalities meet the preservation and incorporation of valuable endoscopic innovations criteria, but they are not necessarily recommended in routine clinical practice [24]. *HD-WLE* high definition white light endoscopy; *4QB* 4 quadrant biopsies according to the Seattle protocol; *NBI* narrow band imaging; *BLI* blue laser imaging; *FICE* Fuji intelligent chromoendoscopy; *iSCAN* is a brand name used by Pentax to refer to their image enhancement technology

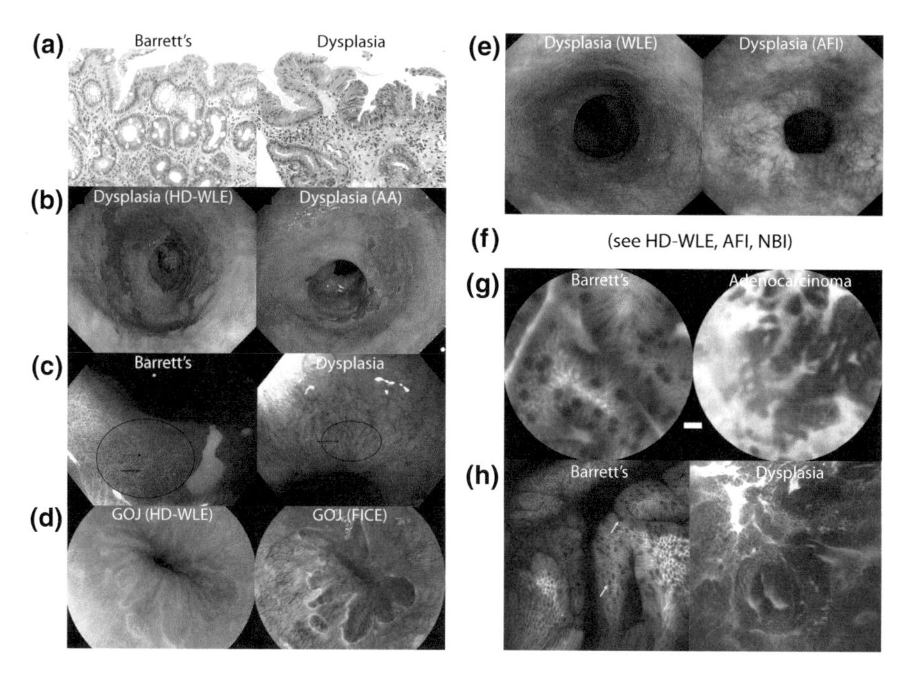

Fig. 2.4 Advanced optical techniques in clinical use for endoscopic surveillance of Barrett's oesophagus—example images. **a** *Histopathology: left*: non-dysplastic Barrett's oesophagus. *Right*: low-grade dysplasia. **b** *Chromoendoscopy, acetic acid: left*: Barrett's with HD-WLE. *Right*: same patient. Note dysplasia only visible post acetic acid (AA) with early loss of acetowhitening (red patch on the lumen in the lower side of the image) (Olympus Lucera ELITE processor, GIFHQ290 gastroscope). Reproduced from [23]. **c** *Hardware-based virtual chromoendoscopy, NBI: left*: Barrett's oesophagus. Note the presence of circular (solid black arrow) and ridge/villous (red arrow) mucosal patterns arranged in an orderly fashion and blood vessels that follow the mucosal ridge architecture (dashed arrows). Reproduced from [30]. *Right*: dysplasia. Note the irregular mucosal (black arrow) and vascular patterns (red arrow). Reproduced from [30]. **D** *Software-based virtual chromoendoscopy, FICE: left*: conventional white-light image of the gastroesophageal junction (GOJ) (from FICE ATLAS, Fujicon). *Right*: Fujicon intelligent chromoendoscopy (FICE) image of the GOJ (From FICE ATLAS Fujicon). Reproduced from [37]. **e** *Autofluorescence imaging: left*: high-grade dysplasia not visible on WLE. Reproduced from [33]. *Right*: AFI reveals high-grade dysplasia as a purple patch. Reproduced from [33]. **F** Endoscopic trimodal imaging: see HD-WLE (**b**), AFI (**e**), NBI (**c**). **g** *pCLE: left*: Barrett's oesophagus. *Right*: early oesophageal adenocarcinoma. Images captured in vivo using GastroFlex UHD (Cellvizio; Mauna Kea Technologies, Paris, France) after injection of sodium fluorescein (2.5 mL, 10%). Scale bar = 20 μm. Reproduced from [38]. **h** *eCLE*: Confocal images of the upper part of the mucosa layer (about 30–50 μm vertical depth). *Left*: Barrett's epithelium. Typical villiform shape and presence of goblet cells (yellow arrows) can be identified. *Right*: high-grade dysplasia. At confocal images black cells with irregular borders and shapes with high dark contrast to surrounding tissue were present. Images captured in vivo with confocal laser endoscope (EC-3870CIFK; Pentax, Tokyo, Japan) after injection of 10% fluorescein sodium (5–10 ml IV). Reproduced from [39]

2.2.2.2 Virtual Chromoendoscopy (Red Flag)

Virtual (also known as electronic or optical) chromoendoscopy improves contrast by modifying the endoscope hardware or software. This avoids the challenges of working with dyes, such as increased procedure time for dye administration and potential for adverse effects caused by the dye (Translational Characteristic 1). Hardware modifications reported to date usually involve adapting the light source to focus on blue and green wavelength bands, where haemoglobin is strongly absorbing, hence providing contrast based on changes in the tissue vasculature [25]. Narrow band imaging (NBI) is the most widely established of these methods and also meets the ASGE thresholds [24]. NBI highlights the mucosal and vascular pattern of the superficial mucosa, enabling the operator to classify the disease state of the tissue based on recognition of altered vascular and mucosal patterns associated with dysplasia [30]. Blue laser imaging (BLI) is a similar technology that has also been tested in patients (in vivo, comparative study, $n = 39$ patients) [42] and is under evaluation in Barrett's oesophagus [43]. Software-based virtual chromoendoscopy methods [44, 45] use proprietary image processing algorithms to improve the contrast of mucosal and surface vessel patterns in the GI tract [37]. While there is currently insufficient data for advisory bodies to make recommendations [24], clinical studies have shown that software-based approaches perform similarly to acetic acid chromoendoscopy (in vivo, prospective randomized pilot study, $n = 57$) [32]. These early findings need to be confirmed with large randomised controlled trials. Virtual chromoendoscopy has significant advantages in being label-free and easily implemented using SOC devices (Translational Characteristic 6) [46–48], so now has widespread availability.

2.2.2.3 Autofluorescence Endoscopy (Red Flag)

Several endogenous structural and metabolic molecules, such as collagen and NADH, are fluorescent. This 'autofluorescence' is reduced in dysplastic tissue compared to the surrounding healthy tissue, providing contrast for autofluorescence imaging (AFI). In commercial endoscopes with AFI, dysplastic tissue is displayed as dark purple patches among a background of green Barrett's fluorescence (Fig. 2.4) [49]. This has high sensitivity for dysplasia, but low specificity since inflammation also reduces tissue autofluorescence [50]. AFI is implemented by adding filters to the light source and detector on a standard endoscope, so it has been combined with HD-WLE as well as virtual chromoendoscopy for endoscopic trimodal imaging (ETMI) in an effort to increase specificity. Trials to date have yielded mixed results [35, 36]; it remains unclear whether AFI truly adds to the already improved performance of NBI. Like NBI, AFI is widely available, potentially due to its incorporation into standard HD-WLE devices (Translational Characteristic 6).

2.2.2.4 Endomicroscopy (Optical Biopsy)

Intravenous fluorescein (a fluorescent dye) can be used to highlight microvasculature and tissue structures. This is commonly examined using confocal laser endomicroscopy (CLE), which produces depth-sectioned, high magnification and high resolution images. These can be used to spot dysplastic changes in cell morphology, for example, irregular vessels or epithelial borders, yielding high sensitivity and specificity for dysplasia [24, 51]. An endoscope-based CLE system (eCLE) was recommended by the ASGE [24], but requires a dedicated endoscope that is no longer on the market. Probe-based CLE (pCLE) uses a thin fibre bundle allowing the probe to be inserted through the working channel of a standard forward facing endoscope. This allows pCLE to occur alongside standard HD-WLE (Translational Characteristic 2) but with lower resolution and limited depth sectioning compared to eCLE. Clinical trial results to date indicate that pCLE can be used to identify neoplasia but is not yet sufficient to replace random biopsies [52].

2.2.3 *Emerging Optical Endoscopic Techniques for Barrett's Surveillance*

The advanced imaging techniques described above are available in clinically approved devices now used in some centres, and three of these techniques, acetic acid chromoendoscopy, NBI and eCLE were recently reported to meet the ASGE performance requirements for targeting biopsy [24]. The eCLE device has since been removed from the market but for NBI and acetic acid chromoendoscopy, this milestone was achieved in part thanks to their having many of the translational characteristics we suggested in Chap. 1. Their deployment at many sites internationally enabled the extensive development of image classification criteria, the publication of consensus statements and the assessment of repeatability and reproducibility (Translational Characteristic 5, Translational Characteristic 4) [30, 40, 53–56]. Both techniques are also compatible with standard forward facing endoscopes, producing familiar 2D images which can be co-registered with HD-WLE (Translational Characteristic 2, Translational Characteristic 6).

These techniques, then, represent the potential to reach widespread implementation by ensuring favourable translational characteristics, as well as performance. Nonetheless, though they are widely used as adjuncts to HD-WLE, neither has yet replaced the standard of care and their recommendation status still hinges on their deployment by expert endoscopists (Translational Characteristic 3, Translational Characteristic 4). Thus, development of novel optical imaging techniques continues, with advances being made that could address outstanding limitations in performance and translatability. These emerging optical techniques are summarised in Table 2.2, with example data from their application in oesophageal tissue shown in Fig. 2.5 [20, 57].

Table 2.2 Emerging optical techniques for endoscopic surveillance of Barrett's oesophagus

Technique	Contrast mechanism	Strengths and weaknesses	Status and prospect in Barrett's oesophagus
Exogenous contrast			
Optical molecular imaging (OMI)	Exogenous fluorophores conjugated to targeting moieties (lectins, peptides, antibodies, affibodies, enzymes) that target intracellular and extracellular proteins and enzymes	+ Specificity − Exogenous contrast − Surface images − Cost	In vivo trials [75] Potential to be translated for wide field surveillance. Awaiting further in vivo trials
Interrogating disordered tissue microstructure			
Optical coherence tomography (OCT)	Structural features cause change of phase which can be used to construct cross sectional image of tissue structure	+ High resolution + Depth sectioning + Endogenous contrast − Large image datasets	In vivo trials (patient series, n = 6) [60]
Elastic scattering spectroscopy (ESS/DRS/LSS)	Endogenous scatterers (organelles, cell nuclei, cell membrane, collagen, cytoplasm, lipids)	+ Depth penetration + Endogenous contrast − Spectrum rather than image	In vivo trials (single centre pilot study, n = 9 patients, n = 95 biopsies) [74]. No trials published in last 10 years
Angle-resolved low coherence interferometry (a/LCI)	Endogenous scatterers (organelles, cell nuclei, cell membrane, collagen, cytoplasm, lipids). Increase in nuclear size with dysplasia can be inferred	+ High sensitivity and specificity in pilot study + Endogenous contrast − Tissue orientation can affect results	In vivo pilot study (2 centre pilot study, n = 46 patients, n = 172 sites) [76]. Combination with OCT. Clinical trials likely
Polarimetry	Microstructural anisotropy	+ Endogenous contrast − Instrumentation challenges	No trials. Awaiting further device development

(continued)

Table 2.2 (continued)

Technique	Contrast mechanism	Strengths and weaknesses	Status and prospect in Barrett's oesophagus
Interrogating abnormal tissue function			
Photoacoustic endoscopy (PAE)	Absorption of endogenous chromophores (melanin, haemoglobin, lipids, water). Allows vascular imaging	+ Volumetric images + Endogenous contrast – Instrumentation challenges – Limited resolution at present – Long acquisition times	In vivo imaging of oesophagus in animals [77]. No trials Awaiting application
Fluorescence lifetime imaging (FLIM)	Endogenous fluorophores (NAD(P)H, flavins, collagen, elastin)	+ More robust than traditional AFI + Endogenous contrast – Safety of UV illumination – Long acquisition times	Ex vivo trials (time-resolved fluorescence, single centre pilot study, n = 37 patients, n = 108 fluorescence decay profiles) [78]. Awaiting in vivo trials
Multi-photon microscopy (MPM)	Endogenous fluorophores (NAD(P)H, flavins, collagen, elastin)	+ Depth sectioning + High resolution + Endogenous contrast – Requires high illumination intensity	Ex vivo trials (n = 25 patients, n = 35 biopsies) [79] Awaiting in vivo trials
Interrogating bulk molecular composition			
Endoscopic Raman spectroscopy (ERS)	Vibrational modes of specific molecules including proteins (proteins, RNA, DNA, lipids, water)	+ Detailed biochemical information + Algorithms have been developed + Multicentre trials underway + Endogenous contrast – Spectrum rather than image – Repeatability has been questioned – Validation using ex vivo tissue is difficult	In vivo trials (pilot study, n = 77 patients) [22] Potential to be translated if repeatability can be confirmed. Awaiting multicentre trials

(continued)

Table 2.2 (continued)

Technique	Contrast mechanism	Strengths and weaknesses	Status and prospect in Barrett's oesophagus
Coherent anti-stokes Raman spectroscopy (CARS)	Specific molecular groups (see ERS)	+ Detailed biochemical information + Increased sensitivity compared to ERS − Instrumentation challenges − Requires high illumination intensity	No trials. Awaiting further device development
Multimodal methods			
Multispectral imaging (MSI)	Endogenous chromophores (melanin, haemoglobin, lipids, water). Endogenous fluorophores (NAD(P)H, flavins, collagen, elastin)	+ Simple + Compact + Endogenous contrast − Surface images	No trials. Awaiting application

Example images for each technique are shown in Fig. 2.5
ESS elastic scattering spectroscopy; *DRS* diffuse reflectance spectroscopy; *LSS* light scattering spectroscopy

2.2.3.1 Interrogating Disordered Tissue Structure

HD-WLE interrogates disordered tissue structure by presenting images of macroscopic abnormalities on the epithelial surface. Recent advances allow endoscopists to probe cross-sectional tissue microstructure, up to several millimetres deep. Optical coherence tomography (OCT) uses scanning low-coherence interferometry to construct cross sectional reflectance images that reveal changes in tissue microstructure arising due to variations in light scattering [58]. Using a set of manual interpretation criteria, such as presence of glands in the images, gives excellent contrast for dysplasia [59, 60]. Wide-field endoscopic applications of OCT were made feasible by the shift from time-domain OCT to optical frequency domain imaging (OFDI), which significantly increased data acquisition rates and allowed 3D images of the entire oesophagus to be acquired using helical scanning inside an inflated balloon to centre the device, a technique often referred to as volumetric laser endomicroscopy (VLE), which is now commercially available [61]. OCT probes may also be housed in tethered capsule endoscopes (Chap. 3), which can be swallowed un-sedated under the supervision of a nurse in primary care, reducing the required expertise (Translational Characteristic 3) [62–64].

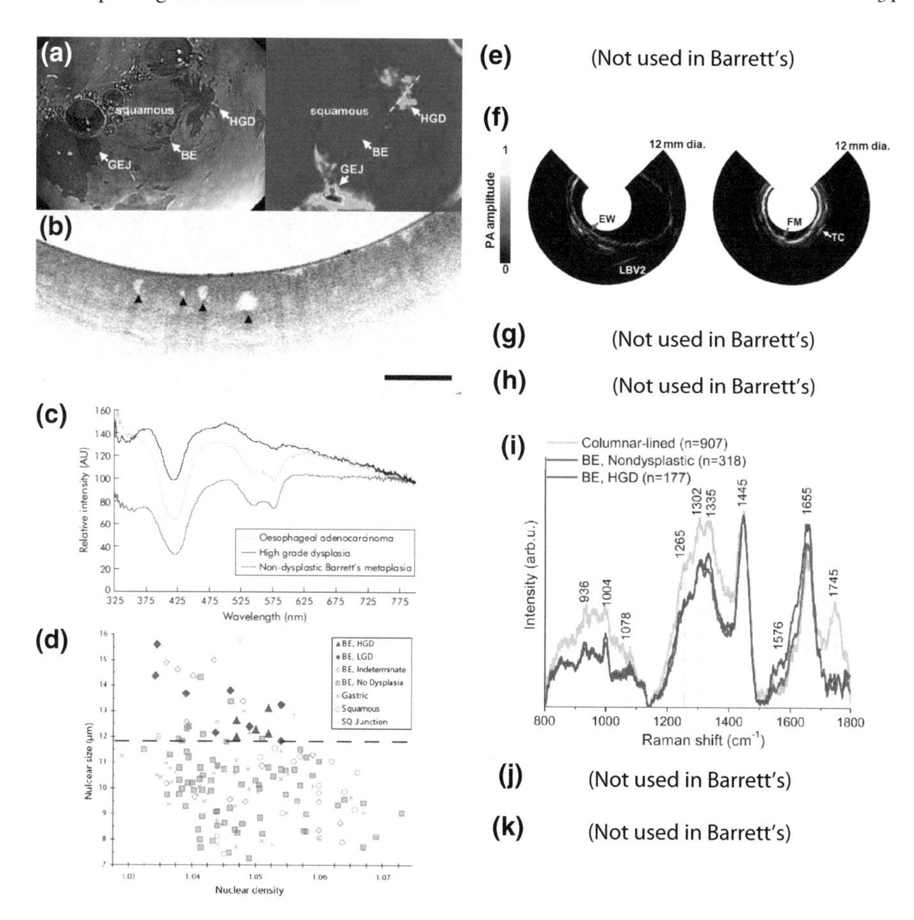

Fig. 2.5 Emerging optical techniques for endoscopic surveillance of Barrett's oesophagus—example images/spectra. **a** *OMI: left*: white-light image showing several areas of Barrett's oesophagus (labelled BE in the diagram), identified by salmon-red mucosa, surrounded by squamous epithelium but no macroscopically visible structural abnormalities to suggest the presence of neoplasia. *Right*: targeted fluorescence image showing enhancement of the signal from HGD. Reproduced from [80]. **b** *OCT*: features consistent with dysplasia, including irregular glandular architecture (arrowheads). Captured with NvisionVLE Imaging System. Scale bar = 1 mm. Reproduced from [81]. **c** *ESS/DRS/LSS*: representative spectra obtained with elastic scattering spectroscopy in vivo. AU: arbitrary units. Reproduced from [70]. **d** *a/LCI*: Scatter plot with each biopsy plotted as a function of its nuclear size and density, as measured by a/LCI system in vivo, and categorised by its pathological diagnosis. Reproduced from [76]. **e** *Polarimetry*: not used in Barrett's. **f** *PAE*: B-scan images of in vivo rabbit oesophagus. Each image covers a 12 mm diameter FOV (~4.1 mm in imaging depth). EW: oesophageal wall, LBV2: large blood vessels, TC: trachea, FM: folded mucosa. Reproduced from [77]. **g** *FLIM*: not used in Barrett's. **h** *MPM*: not used in Barrett's. **i** *ERS*: the mean in vivo confocal Raman spectra of columnar-lined epithelium, non-dysplastic Barrett's, and high-grade dysplasia acquired from Barrett's patients during clinical endoscopy. Each Raman spectrum is acquired within 0.1–0.5 s. Reproduced from [22]. **j** *CARS*: not used in Barrett's. **k** *MSI* Not used in Barrett's

VLE has been successfully correlated with histology in Barrett's patients (ex vivo, feasibility study, n = 14 matched resection specimens) [65] and has been shown to detect oesophageal neoplasia in vivo (in vivo, patient series, n = 6 patients) [60]. Recently, endoscopic OCT has also been used to visualise 3D subsurface microvasculature in the oesophagus [66]. OCT angiography images were interpreted using two microvascular features, abnormal branching and heterogenous vessel size, to allow recognition of dysplasia with 94% sensitivity and 69% specificity (in vivo, n = 52 patients) [67].

A major challenge of OCT-based techniques is that they cannot be used alongside biopsy tools with the existing device architectures (Translational Characteristic 2), so laser cautery marking is being investigated to safely mark regions of interest for later biopsy under HD-WLE guidance [68] (in vivo, pilot study, n = 22 patients). A second challenge remains with image interpretation; an experienced OCT endoscopist is currently needed to interpret images, limiting widespread deployment, but automated image analysis is being investigated to alleviate this burden (Translational Characteristic 4) [69].

In addition to providing contrast for OCT, variation in light absorption and scattering from tissue can be recorded as a function of wavelength or angle. Diffuse reflectance spectroscopy (DRS), also called elastic scattering spectroscopy (ESS), illuminates the tissue with a standard white light source, but instead of collecting an image of the oesophagus using a camera, changes in the spectrum of the light arising from absorption and scattering events in the superficial layers of the tissue are measured with a spectrometer. Contact (ESS) [70] and fixed-distance (DRS) [71] probes can differentiate between healthy and dysplastic tissue in the oesophagus, though are typically restricted to point-based measurements rather than wide-field imaging.

Taking the concept a step further, light scattering spectroscopy (LSS) singles out reflected light that has only scattered once in tissue. The benefit of this approach is that LSS measurements can be directly linked to tissue morphology via physical Mie scattering theory, enabling quantitative measurements of the size and density of cell nuclei, which are typically enlarged and crowded in early cancer [72]. Early LSS studies achieved around 90% sensitivity and specificity for oesophageal dysplasia (in vivo, single centre pilot study, n = 13 patients, n = 76 sample positions) [73] with a contact probe, but unwanted variations in probe-tissue separation led to challenges for interpretation. Hardware developments overcame this limitations to enable 8 cm segments of oesophagus to be mapped with 92% sensitivity and 96% specificity (in vivo, single centre pilot study, n = 9 patients, n = 95 biopsies) [74], showing potential for this to become a useful red-flag tool for guiding targeted biopsies in Barrett's surveillance.

Angle-resolved low coherence interferometry (a/LCI) also looks at singly-scattered light but probes the angular scattering distribution at a single wavelength, from which measurements of nuclear size and nuclear density can be derived. Using these, a/LCI has been shown to identify dysplasia (including LGD) with 100% sensitivity and 84% specificity in vivo (in vivo, 2 centre pilot study, n = 46 patients, n = 172 sample positions) [76] and a negative predictive value of 100%. Although a/LCI was originally a point measurement method, the ability to spatially scan to provide 2D maps that co-register with wide-field images has now been demonstrated [82].

There are also phase and polarisation-sensitive endoscopic methods on the horizon that derive contrast from scattering and present wide-field images [83, 84]. While VLE is currently the most advanced method for interrogating microstructure in terms of clinical translation, if their performance remains high in randomised controlled trials, DRS/ESS, LSS and a/LCI have potential to become valuable optical biopsy techniques, and red flag techniques if 2D spatial scanning can be implemented successfully.

2.2.3.2 Interrogating Abnormal Tissue Function and Metabolism

Angiogenesis, the formation of new blood vessels, is a well-known hallmark of cancer [85], and several optical imaging techniques to probe vascular changes have been investigated. Virtual chromoendoscopy has been successful in improving targeted biopsy based on changes in tissue vasculature, but can only interrogate superficial epithelial changes. OCT has shown exciting possibilities for cross-sectional imaging of tissue function, using measurements of blood flow to highlight the vasculature [86, 87], although these have yet to be clinically validated. Another method that provides cross-sectional vascular information is photoacoustic endoscopy (PAE), which uses optical excitation of tissue to generate ultrasound [88]. Highly optically absorbing molecules in tissue, such as oxy- and deoxy-haemoglobin, can be resolved at far greater penetration depth than is available from exclusively optical imaging, and distinguished based on their absorption spectra, allowing vascular function to be imaged in addition to vascular structure. Thus, PAE could provide high resolution cross-sectional functional imaging at centimetre depths, allowing visualisation of vascular changes associated with dysplasia.

PAE devices using a similar helical scanning implementation to OCT [89–91] have been applied in rabbit oesophagi [77], but further development is needed to increase radial resolution and acquisition speed, as well as to address challenges with image interpretation (similar to those of OCT) (Translational Characteristic 4). Recent successful studies using photoacoustic imaging in breast cancer diagnosis and other areas [92] suggests that application of PAE in Barrett's surveillance may yet yield valuable information, potentially in combination with OCT and other advanced methods [90].

Another well-established marker of dysplasia is the change in tissue autofluorescence associated with underlying changes in metabolism. When measured by standard AFI endoscopes, these changes can be confounded by surface irregularities and non-uniform illumination. Fortunately, measuring the lifetime of the fluorescence signal, rather than its absolute intensity, avoids these confounding factors [93]. Fluorescence lifetime imaging microscopy (FLIM) is able to map changes in local tissue microenvironment and has shown promise in detection of cancers in ex vivo and in vivo studies [93]. For many years, the clinical translation of FLIM was limited by the size, cost and complexity of the instrumentation and the need for long integration times due to weak signals (Translational Characteristic 2). A 2003 study of point-based fluorescence lifetime measurements found sensitivity and specificity for HGD of less than 60% using time resolved fluorescence (in vivo, single centre pilot study, n = 37 patients, n = 108 fluorescence decay profiles) [78]. More recently, however, compact diode-pumped laser-based excitation sources and time-gated methods have addressed instrumentation limitations [94] meaning wide-field FLIM endoscopes with near-video rate acquisition (~2 Hz) [95–97] are now available. While these have been used in vivo [95, 98] they have yet to be applied to Barrett's surveillance and may yield improved performance in this context.

Complementing wide-field FLIM approaches, multi-photon microscopy (MPM) provides an autofluorescence-based alternative to pCLE. MPM is a scanning, optical sectioning, imaging approach in which fluorescence is spatially delineated using non-linear optical excitation. Ex vivo MPM of fresh biopsies can successfully distinguish squamous mucosa, gastric columnar mucosa and intestinal metaplasia (ex vivo, n = 25 patients, n = 35 biopsies) [79] suggesting that MPM could be used to identify dysplasia. MPM endoscopes are being developed and can incorporate additional features from microscopy such as super-resolution imaging [99]. Although MPM is at a very early stage of development, it holds potential to perform high magnification, depth-sectioned, label-free endomicroscopy as part of Barrett's surveillance.

2.2.3.3 Interrogating Bulk Molecular Composition

Changes in bulk molecular composition can be determined using the inelastic scattering spectral 'fingerprint' measured through endoscopic Raman spectroscopy (RS). RS is sensitive to the abundance of molecular bonds, primarily lipid, protein and nucleic acid content in tissue. RS data is typically classified using machine-learning methods that use a training set of spectra where the disease classification is known from histopathology analysis [100].

Historically, the low intensity of Raman signals has been a hurdle for endoscopic RS, resulting in very slow data acquisition (Translational Characteristic 2). Nonetheless, Bergholt et al. recently demonstrated an ERS probe, including a classification algorithm based on a Raman library of >12,000 spectra, that could differentiate between columnar lined epithelium, non-dysplastic Barrett's or HGD, in real time (0.2 s), passing this information to the endoscopist using auditory feedback as the probe was passed across the tissue (in vivo, pilot study, n = 77 patients, sensitivity 87.0%, specificity 84.7%) (Translational Characteristic 4) [22]. Alternatively, coherent anti-Stokes Raman spectroscopy (CARS), which uses multiple photons to probe specific regions of the spectral fingerprint, has been suggested as a way to overcome the low Raman intensity. Several prototype endoscopes have been developed, despite challenges associated with non-linear effects in fibres and the design of miniature optics [101, 102]. Precision remains a challenge for both endoscopic RS and CARS, due to pressure-based signal variation, but prospective randomized multicentre trials are underway [22] and further devices are in development [103]. Raman spectroscopy is thus a promising technique that could provide both point measurements for optical biopsy, and 2D scanning measurements to assess larger areas of tissue for red flag surveillance.

2.2.3.4 Multimodal Methods

The recent advances in optical techniques highlighted above suggest that contrast mechanisms based on intrinsic optical interactions with tissue have the potential to improve the diagnostic yield of Barrett's surveillance. Naturally, combining several of these contrast mechanisms into a single device could have added benefits, giving access to structural, functional and molecular information simultaneously. For example, a recent pilot study combining intraoperative DRS, RS and fluorescence spectroscopy achieved 100% sensitivity and 93% specificity for several cancers (in vivo, $n = 15$ patients) [104] and another study combining fluorescence spectroscopy, DRS and LSS achieved 93% sensitivity and 100% specificity for dysplasia in Barrett's oesophagus (in vivo, $n = 16$ patients) [105]. Achieving a successful combination, however, requires careful optical design and often complex instrumentation (Translational Characteristic 2).

One promising route to overcoming this challenge may lie in the use of multispectral imaging (MSI), where the light illuminating the tissue and being detected is split into its component colours, or rapidly modulated. NBI is a simple example of this, where just 2 wavelength bands of illumination are used, and standard WLE another, where just 3 wavelength bands are detected. MSI goes further, recording 10s or 100s of colours at every pixel in an endoscopic image, which can then be processed to spatially resolve reflectance (e.g. NBI), fluorescence (e.g. AFI) or Raman information, depending on the illumination applied and the signal detected. MSI has already shown potential for aiding cancer diagnosis in a range of organ sites, including the oesophagus [106], although its potential has yet to be demonstrated through *in vivo* trials in Barrett's surveillance. MSI hardware is often bulky and slow (Translational Characteristic 2) but optimisation of the MSI hardware, for example using compact spectrally resolved detector arrays (SRDAs) [107], may assist with real-time clinical application in the future.

2.2.3.5 Optical Molecular Imaging

Despite the aforementioned disadvantages of exogenous contrast, application of exogenous molecular imaging agents consisting of a targeting probe and fluorescent optical reporter has the potential to allow wide field visualisation of complex biochemical processes involved in both normal physiology and disease with unparalleled specificity. Optical molecular imaging (OMI) has the potential to allow wide-field red flag visualisation of biochemical changes associated with dysplasia. Molecular imaging agents may be administered topically or intravenously. Intravenous administration uses lower concentrations of probe and is able to target entire tumours rather than only the exposed surfaces, but requires long (hours-days) incubation times. Topical application is fast (minutes) and limits bio-distribution of the probe to the organ of interest, limiting toxicity concerns. In Barrett's surveillance, topical application is preferred as it causes minimum interruption to the endoscopy workflow whilst still reaching the surface cells of interest. Near-infrared rather than visible fluorophores are preferred in order to avoid confounding visible autofluorescence background.

Due to the heterogeneity of gene expression in Barrett's, it is likely that multiple molecular imaging agents will be required to reach high sensitivity and specificity. Since each of these requires a lengthy toxicology analysis before *in vivo* validation can even begin, translation of OMI is slow (Translational Characteristic 1), but the enticing promise of macroscopic imaging of microscopic biochemical changes, and encouraging results in current trials, continues to drive much development (including our work in Chap. 4).

2.3 Summary

Though several advanced endoscopy techniques have met the ASGE performance standards for recommended routine use in Barrett's oesophagus, histological assessment of HD-WLE targeted and random biopsies is still the standard of care. Many emerging techniques achieve some of the aforementioned requirements, but fall short of providing the ideal solution. Continuing advances in hardware and software are allowing endoscopic application of optical contrast mechanisms developed for other indications. It is likely that the next decade will see some of these exciting new contrast mechanisms applied to Barrett's surveillance in new devices for the first time, potentially leading to techniques that provide a long-awaited improvement to the standard of care.

Reviewing the optical techniques presented in this chapter, it can be concluded that an ideal endoscopic surveillance method should perform comprehensive investigation of the oesophagus with high sensitivity and specificity for dysplasia. It should allow for the use of endoscopic tools for marking and biopsy if necessary (Translational Characteristic 2). If exogenous contrast is chosen, it should be topically applied and provide contrast enhancement enough to justify the increased procedure times and costs, which should be minimised (Translational Characteristic 1). Implementation of the technique should be possible with minimal additional training of endoscopist operators and image interpreters (Translational Characteristic 3, Translational Characteristic 4). To achieve widespread deployment in healthcare systems, no significant change to procedure times, expertise or costs should be made; ideally these would be reduced. Though initial trials are facilitated by the device being compatible with the current standard of care (Translational Characteristic 2, Translational Characteristic 6), if possible, availability of endoscopy to patients should be increased, for example by enabling deployment by non-specialist operators in primary care centres, and physical discomfort with the procedure should be decreased, to improve compliance with surveillance programmes.

From this review of optical techniques, it is evident that optical molecular imaging (OMI) and multispectral imaging (MSI) provide potentially translatable solutions in Barrett's surveillance, satisfying many of the aforementioned criteria: they provide wide-field 2D images (Translational Characteristic 3), allowing red flag surveillance and coregistration with familiar 2D HD-WLE images (Translational Characteristic 6); they are easily implemented in a standard forward facing device architecture (Translational Characteristic 2); and they acquire images that can potentially be processed to generate a simple intensity map, with bright pixels highlighting suspicious regions of dysplasia, allowing easy interpretation without the need for complex classification algorithms (Translational Characteristic 4). Additionally, MSI utilises endogenous contrast (Translational Characteristic 1).

Over the next three chapters, work translating two novel techniques towards clinical implementation is described. First, the choice of device architecture under which the prototype imaging devices were prepared is explained, including its advantages and limitations. This is important, as it simplified approval of the devices for use in humans and facilitated subsequent deployment in clinical trials. However, this device architecture introduced artefacts into the images, so the remainder of Chap. 3 is concerned with outlining the image correction algorithms developed to remove these. Next, Chap. 4 describes work on OMI. The design and characterisation of an endoscope for NIR fluorescence imaging of a topically applied molecular imaging agent that binds specifically to Barrett's oesophagus and not to dysplasia is described, and the validation of this technique using a range of biological samples is presented. Finally, in Chap. 5, work developing an endoscope for MSI of endogenous tissue contrast is described. This work led to a pilot clinical trial of multispectral endoscopy in Barrett's surveillance.

References

1. CRUK. Resources|CRUK Cambridge Centre early detection programme. Available at: https://www.earlydetectioncambridge.org.uk/resources. Accessed 18 Jan 2018
2. Gatenby P et al (2014) Lifetime risk of esophageal adenocarcinoma in patients with Barrett's esophagus. World J Gastroenterol WJG 20:9611–9617
3. Desai TK et al (2012) The incidence of oesophageal adenocarcinoma in non-dysplastic Barrett's oesophagus: a meta-analysis. Gut 61:970–976
4. Bhat S et al (2011) Risk of malignant progression in Barrett's esophagus patients: results from a large population-based study. J Natl Cancer Inst 103:1049–1057
5. Hvid-Jensen F, Pedersen L, Drewes AM, Sørensen HT, Funch-Jensen P (2011) Incidence of adenocarcinoma among patients with Barrett's esophagus. New Engl J Med 365:1375–1383
6. Duits LC et al (2015) Barrett's oesophagus patients with low-grade dysplasia can be accurately risk-stratified after histological review by an expert pathology panel. Gut 64:700–706
7. Reid BJ, Levine DS, Longton G, Blount PL, Rabinovitch PS (2000) Predictors of progression to cancer in Barrett's esophagus: baseline histology and flow cytometry identify low- and high-risk patient subsets. Am J Gastroenterol 95:1669–1676
8. CRUK. Oesophageal cancer statistics. http://www.cancerresearchuk.org/health-professional/cancer-statistics/statistics-by-cancer-type/oesophageal-cancer

9. Barbour AP et al (2010) Risk stratification for early esophageal adenocarcinoma: analysis of lymphatic spread and prognostic factors. Ann Surg Oncol 17:2494–2502
10. Fitzgerald RC et al (2014) British Society of Gastroenterology guidelines on the diagnosis and management of Barrett's oesophagus. Gut 63:7–42
11. Weusten BLAM et al (2017) Endoscopic management of Barrett's esophagus: European Society of Gastrointestinal Endoscopy (ESGE) Position Statement. Endoscopy 191–198 (2017)
12. Shaheen NJ, Falk GW, Iyer PG, Gerson LB, American College of Gastroenterology (2016) ACG clinical guideline: diagnosis and management of Barrett's esophagus. Am J Gastroenterol 111:30–50
13. Evans JA et al (2012) The role of endoscopy in Barrett's esophagus and other premalignant conditions of the esophagus. Gastrointest Endosc 76:1087–1094
14. Spechler SJ, Sharma P, Souza RF, Inadomi JM, Shaheen NJ (2011) American Gastroenterological Association technical review on the management of Barrett's esophagus. Gastroenterology 140:e18–e52
15. Kastelein F et al (2015) Impact of surveillance for Barrett's oesophagus on tumour stage and survival of patients with neoplastic progression. Gut 65:1–7
16. Verbeek RE et al (2014) Surveillance of Barrett's esophagus and mortality from esophageal adenocarcinoma: a population-based cohort study. Am J Gastroenterol 109:1215–1222
17. El-Serag HB et al (2016) Surveillance endoscopy is associated with improved outcomes of oesophageal adenocarcinoma detected in patients with Barrett's oesophagus. Gut 65:1252–1260
18. Corley DA et al (2013) Impact of endoscopic surveillance on mortality from Barrett's esophagus-associated esophageal adenocarcinomas. Gastroenterology 145:312–319
19. Levine DS, Blount PL, Rudolph RE, Reid BJ (2000) Safety of a systematic endoscopic biopsy protocol in patients with Barrett's esophagus. Am J Gastroenterol 95:1152–1157
20. Sturm MB, Wang TD (2015) Emerging optical methods for surveillance of Barrett's oesophagus. Gut 64:1816–1823
21. Brown H et al (2015) Scoping the future: an evaluation of endoscopy capacity across the NHS in England
22. Bergholt MS et al (2014) Fiberoptic confocal raman spectroscopy for real-time in vivo diagnosis of dysplasia in Barrett's esophagus. Gastroenterology 146:27–32
23. Chedgy FJQ, Subramaniam S, Kandiah K, Thayalasekaran S, Bhandari P (2016) Acetic acid chromoendoscopy: improving neoplasia detection in Barrett's esophagus. World J Gastroenterol 22:5753–5760
24. Thosani N et al (2016) ASGE Technology Committee systematic review and meta-analysis assessing the ASGE preservation and incorporation of valuable endoscopic innovations thresholds for adopting real-time imaging-assisted endoscopic targeted biopsy during endoscopic surveillance. Gastrointest Endosc 83:684–698
25. Beg S, Wilson A, Ragunath K (2016) The use of optical imaging techniques in the gastrointestinal tract. Frontline Gastroenterol 7:207–215
26. Swager A, Curvers WL, Bergman JJ (2015) Diagnosis by endoscopy and advanced imaging. Best Pract Res Clin Gastroenterol 29:97–111
27. Olliver JR, Wild CP, Sahay P, Dexter S, Hardie LJ (2003) Chromoendoscopy with methylene blue and associated DNA damage in Barrett's oesophagus. Lancet 362:373–374
28. Coletta M et al (2016) Acetic acid chromoendoscopy for the diagnosis of early neoplasia and specialized intestinal metaplasia in Barrett's esophagus: a meta-analysis. Gastrointest Endosc 83:57–67
29. Sharma P et al (2013) Standard endoscopy with random biopsies versus narrow band imaging targeted biopsies in Barrett's oesophagus: a prospective, international, randomised controlled trial. Gut 62:15–21
30. Sharma P et al (2016) Development and validation of a classification system to identify high-grade dysplasia and esophageal adenocarcinoma in Barrett's esophagus using narrow band imaging. Gastroenterology 150:591–598

31. Maes S, Sharma P, Bisschops R (2016) Review: surveillance of patients with Barrett oesophagus. Best Pract Res Clin Gastroenterol 30:901–912
32. Pohl J et al (2007) Comparison of computed virtual chromoendoscopy and conventional chromoendoscopy with acetic acid for detection of neoplasia in Barrett's esophagus. Endoscopy 39:594–598
33. Boerwinkel DF et al (2014) Effects of autofluorescence imaging on detection and treatment of early neoplasia in patients with Barrett's esophagus. Clin Gastroenterol Hepatol 12:774–781
34. Giacchino M et al (2013) Clinical utility and interobserver agreement of autofluorescence imaging and magnification narrow-band imaging for the evaluation of Barrett's esophagus: a prospective tandem study. Gastrointest Endosc 77:711–718
35. Curvers WL et al (2010) Endoscopic tri-modal imaging is more effective than standard endoscopy in identifying early-stage neoplasia in Barrett's esophagus. Gastroenterology 139:1106–1114
36. Curvers WL et al (2011) Endoscopic trimodal imaging versus standard video endoscopy for detection of early Barrett's neoplasia: a multicenter, randomized, crossover study in general practice. Gastrointest Endosc 73:195–203
37. Manfredi MA et al (2015) Electronic chromoendoscopy. Gastrointest Endosc 81:249–261
38. Sharma P et al (2011) Real-time increased detection of neoplastic tissue in Barrett's esophagus with probe-based confocal laser endomicroscopy: final results of an international multicenter, prospective, randomized, controlled trial. Gastrointest Endosc 74:465–472
39. Trovato C et al (2013) Confocal laser endomicroscopy for in vivo diagnosis of Barrett's oesophagus and associated neoplasia: a pilot study conducted in a single Italian centre. Dig Liver Dis 45:396–402
40. Longcroft-Wheaton G et al (2013) Duration of acetowhitening as a novel objective tool for diagnosing high risk neoplasia in Barrett's esophagus: a prospective cohort trial. Endoscopy 45:426–432
41. Ngamruengphong S, Sharma VK, Das A (2009) Diagnostic yield of methylene blue chromoendoscopy for detecting specialized intestinal metaplasia and dysplasia in Barrett's esophagus: a meta-analysis. Gastrointest Endosc 69:1021–1028
42. Kaneko K et al (2014) Effect of novel bright image enhanced endoscopy using blue laser imaging (BLI). Endosc Int Open 02:E212–E219
43. Osawa H et al (2014) Blue laser imaging provides excellent endoscopic images of upper gastrointestinal lesions. Video J Encycl GI Endosc 1:607–610
44. Miyake Y et al (2005) Development of new electronic endoscopes using the spectral images of an internal organ. In: Proceedings of the IS&T/SID's thirteen color imaging conference, Society for Imaging Science and Technology, pp 261–269
45. Kodashima S, Fujishiro M (2010) Novel image-enhanced endoscopy with i-scan technology. World J Gastroenterol 16:1043–1049
46. Rodriguez SA et al (2010) Ultrathin endoscopes. Gastrointest Endosc 71:893–898
47. Imagawa H et al (2011) Improved visibility of lesions of the small intestine via capsule endoscopy with computed virtual chromoendoscopy. Gastrointest Endosc 73:299–306
48. Dung LR, Wu YY (2010) A wireless narrowband imaging chip for capsule endoscope. IEEE Trans Biomed Circuits Syst 4:462–468
49. von Holstein CS et al (1996) Detection of adenocarcinoma in Barrett's oesophagus by means of laser induced fluorescence. Gut 39:711–716
50. Kara MA et al (2005) Endoscopic video autofluorescence imaging may improve the detection of early neoplasia in patients with Barrett's esophagus. Gastrointest Endosc 61:679–685
51. Wallace M et al (2011) Miami classification for probe-based confocal laser endomicroscopy. Endoscopy 43:882–891
52. Xiong YQ, Ma SJ, Zhou JH, Zhong XS, Chen Q (2016) A meta-analysis of confocal laser endomicroscopy for the detection of neoplasia in patients with Barrett's esophagus. J Gastroenterol Hepatol (Australia) 31:1102–1110
53. Kara MA, Ennahachi M, Fockens P, ten Kate FJW, Bergman JJGHM (2006) Detection and classification of the mucosal and vascular patterns (mucosal morphology) in Barrett's esophagus by using narrow band imaging. Gastrointest Endosc 64:155–166

54. Singh R et al (2008) Narrow-band imaging with magnification in Barrett's esophagus: validation of a simplified grading system of mucosal morphology patterns against histology. Endoscopy 40:457–463
55. Sharma P et al (2006) The utility of a novel narrow band imaging endoscopy system in patients with Barrett's esophagus. Gastrointest Endosc 64:167–175
56. Kandiah K et al (2016) OC-054 development and validation of a classification system to identify Barrett's neoplasia using acetic acid chromoendoscopy: the predict classification: Abstract OC-054 Table 1. Gut 65:A31.1–A31 (2016)
57. Robles LY, Singh S, Fisichella PM (2015) Emerging enhanced imaging technologies of the esophagus: spectroscopy, confocal laser endomicroscopy, and optical coherence tomography. J Surg Res 195:502–514
58. Gora MJ, Suter MJ, Tearney GJ, Li X (2017) Endoscopic optical coherence tomography: technologies and clinical applications [Invited]. Biomed Opt Express 8:2405
59. Leggett CL et al (2015) Comparative diagnostic performance of volumetric laser endomicroscopy and confocal laser endomicroscopy in the detection of dysplasia associated with Barrett's esophagus. Gastrointest Endosc 83:880–888.e2
60. Trindade AJ, George BJ, Berkowitz J, Sejpal DV, McKinley MJ (2016) Volumetric laser endomicroscopy can target neoplasia not detected by conventional endoscopic measures in long segment Barrett's esophagus. Endosc Int Open 4:E318–E322
61. NvisionVLE® Imaging System—NinePoint Medical. Available at: http://www.ninepointmedical.com/nvisionvle-imaging-system/. Accessed 1st Aug 2017
62. Gora MJ et al (2013) Tethered capsule endomicroscopy enables less invasive imaging of gastrointestinal tract microstructure. Nat Med 19:238–240
63. Gora MJ et al (2013) Imaging the upper gastrointestinal tract in unsedated patients using tethered capsule endomicroscopy. Gastroenterology 145:723–725
64. Gora M et al (2016) Tethered capsule endomicroscopy: from bench to bedside at a primary care practice Tethered capsule endomicroscopy: from bench to bedside at a primary care practice. J Biomed Opt 21:104001
65. Swager A et al (2015) Volumetric laser endomicroscopy in Barrett's esophagus: a feasibility study on histological correlation. Dis Esophagus 1–8 (2015)
66. Tsai TH et al (2014) Endoscopic optical coherence angiography enables 3-dimensional visualization of subsurface microvasculature. Gastroenterology 147:1219–1221
67. Lee HC et al (2017) Endoscopic optical coherence tomography angiography microvascular features associated with dysplasia in Barrett's esophagus (with video). Gastrointest Endosc 86:476–484
68. Suter MJ et al (2014) Esophageal-guided biopsy with volumetric laser endomicroscopy and laser cautery marking: a pilot clinical study. Gastrointest Endosc 79:886–896
69. Ughi GJ et al (2016) Automated segmentation and characterization of esophageal wall in vivo by tethered capsule optical coherence tomography endomicroscopy. Biomed Opt Express 7:660–665
70. Lovat LB et al (2006) Elastic scattering spectroscopy accurately detects high grade dysplasia and cancer in Barrett's oesophagus. Gut 55:1078–1083
71. Douplik A et al (2014) Diffuse reflectance spectroscopy in Barrett's esophagus: developing a large field-of-view screening method discriminating dysplasia from metaplasia. J Biophotonics 7:304–311
72. Perelman LT, Backman V (2016) Light scattering spectroscopy of epithelial tissue: principles and applications. In: Handbook of optical biomedical diagnostics. SPIE PRESS
73. Wallace M et al (2000) Endoscopic detection of dysplasia in patients with Barrett's esophagus using light-scattering spectroscopy. Gastroenterology 119:677–682
74. Qiu L et al (2012) Spectral imaging with scattered light: from early cancer detection to cell biology. IEEE J Sel Top Quantum Electron 18:1073–1083
75. Lee JH, Wang TD (2016) Molecular endoscopy for targeted imaging in the digestive tract. Lancet Gastroenterol Hepatol 1:147–155

76. Terry NG et al (2011) Detection of dysplasia in Barrett's esophagus with in vivo depth-resolved nuclear morphology measurements. Gastroenterology 140:42–50
77. Yang JM et al (2015) Three-dimensional photoacoustic and ultrasonic endoscopic imaging of two rabbit esophagi. Proc SPIE 9323
78. Pfefer TJ, Paithankar DY, Poneros JM, Schomacker KT, Nishioka NS (2003) Temporally and spectrally resolved fluorescence spectroscopy for the detection of high grade dysplasia in Barrett's esophagus. Lasers Surg Med 32:10–16
79. Chen J, Wong S, Nathanson MH, Jain D (2014) Evaluation of Barrett esophagus by multi-photon microscopy. Arch Pathol Lab Med 138:204–212
80. Joshi BP et al (2016) Multimodal endoscope can quantify wide-field fluorescence detection of Barrett's neoplasia. Endoscopy 48
81. Wolfsen HC et al (2015) Safety and feasibility of volumetric laser endomicroscopy in patients with Barrett's esophagus (with videos). Gastrointest Endosc 82:631–640
82. Kim S et al (2016) Analyzing spatial correlations in tissue using angle-resolved low coherence interferometry measurements guided by co-located optical coherence tomography. Biomed Opt Express 7:1400
83. Qi J, Elson DS (2016) A high definition Mueller polarimetric endoscope for tissue characterisation. Sci Rep 6:25953
84. Ba C, Palmiere M, Ritt J, Mertz J (2016) Dual-modality endomicroscopy with co-registered fluorescence and phase contrast. Biomed Opt Express 7:3403
85. Hanahan D, Weinberg RA (2011) Hallmarks of cancer: the next generation. Cell 144:646–674
86. Tan ACS et al (2017) An overview of the clinical applications of optical coherence tomography angiography. Eye 1–25 (2017)
87. Kashani AH et al (2017) Optical coherence tomography angiography: a comprehensive review of current methods and clinical applications. Prog Retinal Eye Res 60:66–100
88. Wang LV, Yao J (2016) A practical guide to photoacoustic tomography in the life sciences. Nat Methods 13:627–638
89. Yang J-M et al (2015) Optical-resolution photoacoustic endomicroscopy in vivo. Biomed Opt Express 6:918
90. Dong B, Chen S, Zhang Z, Sun C, Zhang HF (2014) Photoacoustic probe using a microring resonator ultrasonic sensor for endoscopic applications. Opt Lett 39:4372–4375
91. Bai X et al (2014) Intravascular optical-resolution photoacoustic tomography with a 1.1 mm diameter catheter. PLOS ONE 9:e92463
92. Zackrisson S, van de Ven SMWY, Gambhir SS (2014) Light in and sound out: emerging translational strategies for photoacoustic imaging. Can Res 74:979–1004
93. Marcu L (2012) Fluorescence lifetime techniques in medical applications. Ann Biomed Eng 40:304–331
94. McGinty J et al (2010) Wide-field fluorescence lifetime imaging of cancer. Biomed Opt Express 1:627–640
95. Sun Y et al (2010) Fluorescence lifetime imaging microscopy for brain tumor image-guided surgery. J Biomed Opt 15:056022
96. Cheng S et al (2013) Flexible endoscope for continuous in vivo multispectral fluorescence lifetime imaging. Opt Lett 38:1515–1517
97. Sparks H et al (2015) A flexible wide-field FLIM endoscope utilising blue excitation light for label-free contrast of tissue. J Biophotonics 8:168–178
98. Sun Y et al (2013) Endoscopic fluorescence lifetime imaging for in vivo intraoperative diagnosis of oral carcinoma. Microsc Microanal (Official Journal of Microscopy Society of America, Microbeam Analysis Society, Microscopical Society of Canada) 19:791–798
99. Gu M, Kang H, Li X (2014) Breaking the diffraction-limited resolution barrier in fiber-optical two-photon fluorescence endoscopy by an azimuthally-polarized beam. Sci Rep 4:3627
100. Jermyn M et al (2016) A review of Raman spectroscopy advances with an emphasis on clinical translation challenges in oncology. Phys Med Biol 61:R370–R400
101. Wang Z et al (2011) Use of multimode optical fibers for fiber-based coherent anti-stokes Raman scattering microendoscopy imaging. Opt Lett 36:2967–2969

102. Légaré F, Evans CL, Ganikhanov F, Xie XS (2006) Towards CARS endoscopy. Opt Express 14:4427–4432
103. Almond LM et al (2014) Endoscopic Raman spectroscopy enables objective diagnosis of dysplasia in Barrett's esophagus. Gastrointest Endosc 79:37–45
104. Jermyn M et al (2017) Highly accurate detection of cancer in situ with intraoperative, label-free, multimodal optical spectroscopy. Can Res 77:3942–3950
105. Georgakoudi I et al (2001) Fluorescence, reflectance, and light-scattering spectroscopy for evaluating dysplasia in patients with Barrett's esophagus. Gastroenterology 120:1620–1629
106. Lu G, Fei B (2014) Medical hyperspectral imaging: a review. J Biomed Opt 19:10901
107. Luthman AS, Dumitru S, Quiros-Gonzalez I, Joseph J, Bohndiek SE (2017) Fluorescence hyperspectral imaging (fHSI) using a spectrally resolved detector array. J Biophotonics 10:840–853

Chapter 3
Flexible Endoscopy: Device Architecture

In order to apply a novel optical imaging technique, a contrast mechanism must be combined with a device architecture (Table 3.1). Here, emerging endoscopic device architectures are described, and their impact on clinical translation is considered. This is followed by a description of the device architecture we chose to develop, the work done to address its limitations, and the advantages this device architecture affords us.

3.1 Flexible Endoscopic Device Architectures

At present, clinical endoscopic surveillance of Barrett's oesophagus is performed using a forward-facing trans-oral endoscope architecture (Fig. 3.1a), around which standard endoscopic tools, such as biopsy forceps and treatment devices, have been designed. Often these tools access the patient via a hollow working channel in the body of the endoscope. Forward-facing endoscopes require articulation by the endoscopist to ensure complete surveillance of the tissue. A key consideration in the development and clinical translation of new optical techniques is whether this remains the appropriate device architecture. An obvious approach to advanced endoscopy is simply to integrate the new contrast mechanism into an existing forward-facing endoscope, to exploit familiarity of endoscopists with the device and retain access to the usual endoscopic tools. In recent years, research groups and commercial companies have taken a more 'out-of-the-box' approach, developing a host of alternative device architectures that overcome some or all of the limitations of forward-facing endoscopy, namely: low magnification; high procedure cost; the need for specialist operators; and restricted angular field of view (Table 3.1).

Accessory channel endoscopes (Fig. 3.1b), are small-diameter probes that can be inserted into standard forward-facing endoscopes through the channel that is normally used to introduce tools. This allows them to be controlled by the standard endoscope; allows their images to be coregistered with standard imaging; and allows them to be implemented without overhaul of the existing equipment and facilities

D. J. Waterhouse, *Novel Optical Endoscopes for Early Cancer Diagnosis and Therapy*, Springer Theses, https://doi.org/10.1007/978-3-030-21481-4_3

Table 3.1 Endoscopic device architectures

Device architecture	Example device(s)	Field of view	Advantages	Disadvantages	Most compatible contrast mechanisms[a]	Typical image types[b]
Forward facing (trans-oral)	Standard commercial endoscopes e.g. Olympus, Pentax, Fuji	Wide (typically 140° luminal view)	Availability and familiarity; wide variety of tools for biopsy, washing, marking; articulation	Endoscopist must articulate to survey entire surface	WLE, Chromoendoscopy, NBI, eCLE, OMI, MSI, Polarimetry	En face, luminal
Forward facing (trans-nasal)	Standard commercial endoscopes e.g. Olympus, Pentax, Fuji	Wide (typically 140° luminal view)	Improved patient tolerance and no sedation required; articulation; shorter, less costly procedure	Endoscopist must articulate to survey entire surface; lower quality image; narrower working channel inappropriate for interventions; poorer suction and air function and smaller biopsy capabilities compared with trans-nasal endoscopes. Unsuitable for Barrett's surveillance	WLE, NBI	En face, luminal
Accessory channel (e.g. contact probe)	Mauna Kea Cellvizio®	Narrow for contact probe (10s–100s microns)	Compatible with insertion through working channel of standard endoscopes	Must be used alongside standard endoscope for articulation, washing, biopsy, marking; contact with lumen must be carefully controlled; small field of view	pCLE, ERS, ESS/DRS, a/LCI, MSI, FLIM, MPM, PA, Polarimetry	Spectral, en face

(continued)

Table 3.1 (continued)

Device architecture	Example device(s)	Field of view	Advantages	Disadvantages	Most compatible contrast mechanisms[a]	Typical image types[b]
Balloon based	NinePoint Nvision-VLE®	Volumetric	Controlled withdrawal; potential for cautery marking capability; compatible with insertion through working channel of standard endoscopes; allows full volumetric imaging of oesophagus	No biopsy or washing capabilities	OCT	Volumetric
Wireless capsule	Given imaging PillCam® ESO series	2 × 169° (ESO2)	No sedation required; can be implemented in primary care; potential for low cost if reusable	One shot (cannot return to suspicious lesions); no biopsy, washing, marking capabilities; long delay for capsule to pass (8–10 h); no control over motion; contact with lumen must be controlled	WLE, NBI, MSI, Polarimetry	En face, luminal or circumfer-ential
Tethered capsule	No commercial devices	Volumetric	No sedation required; can be implemented in primary care; potential for low cost if reusable; controlled withdrawal; potential for cautery marking capability; immediate removal of capsule; allows full volumetric imaging of oesophagus	No biopsy, washing capabilities; contact with lumen must be controlled	OCT	Volumetric

(continued)

Table 3.1 (continued)

Device architecture	Example device(s)	Field of view	Advantages	Disadvantages	Most compatible contrast mechanisms[a]	Typical image types[b]
Wide angle	EndoChoice Fuse (330°)	Extra wide (>140°)	Familiarity; wide variety of tools for biopsy, washing, marking; articulation; wide field of view allows viewing of entire lumen with minimal articulation	Increased cost	WLE, Chromoendoscopy, NBI, eCLE, OMI, MSI, Polarimetry	En face, circumferential

Schematics of each device architecture are shown in Fig. 3.1

[a]Theoretically most combinations of contrast mechanism and device type are possible. Here we give the contrast mechanisms that are most compatible with the advantages and disadvantages of the device architecture

[b]Image type is again dependent on contrast mechanism. Here we give the image types for the most compatible contrast mechanisms for the device architecture

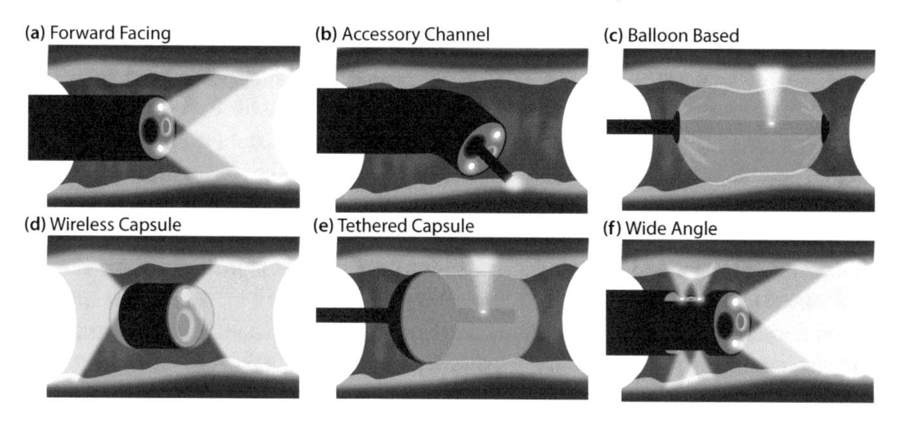

Fig. 3.1 Endoscopic device architectures—schematics. Schematic representations of the endoscopic device architectures listed in Table 3.1

(Translational Characteristic 6). In contrast to forward-facing trans-oral endoscopes, where the camera is found on the tip of the endoscope, accessory channel endoscopes often relay light to detectors outside the patient, using a fibre bundle consisting of 10,000–50,000 individual glass fibrelets [1].

These probes are often placed in direct contact with oesophageal tissue to perform optical biopsy. Unlike physical biopsy, increasing the number of these optical biopsies does not add significantly to procedure cost, however, the endoscopist must manually control the position of the accessory channel endoscope using the standard endoscope, so the results will still be subject to sampling error. In addition to this, the physiological movements of the oesophagus due to peristalsis and anatomic vicinity to the heart, make stabilisation of probe-tissue contact challenging. In other cases, rather than direct contact, the imaging device employs a balloon (Fig. 3.1c) that is inflated to ensure a fixed distance between the tissue and the central axis of the imaging hardware [2, 3].

The high procedure cost of forward-facing endoscopy arises from the need for patient sedation in a specialist facility with a skilled endoscopist. Un-sedated trans-nasal endoscopy (UTNE) provides standard endoscopy capabilities (imaging, articulation, insufflation, suction, biopsy) in a slim device that can be used without sedation, as the trans-nasal intubation does not involve contact with the root of the tongue and therefore does not trigger the gagging reflex. Multiple UTNE systems are commercially available, including two disposable devices which make reprocessing feasible outside of a hospital environment [4]. UTNE has been successfully used for imaging Barrett's and dysplasia [5–7]. Recent studies using UTNE to screen for Barrett's have found it to be comparable to standard endoscopy in clinical effectiveness, participation and safety [8] and considerably cheaper, especially if implemented in a mobile unit instead of a hospital [9]. Nonetheless, UTNE image quality is currently insufficient for dysplasia detection.

While UTNE still requires a skilled endoscopist, wireless capsule endoscopes (Fig. 3.1d) are single-use, pill-shaped devices that can be administered by a non-specialist operator (Translational Characteristic 3). Originally developed for small bowel imaging [10], wireless capsules have since been developed for the oesophagus [11]. When surveyed, most patients prefer capsule endoscopy to regular endoscopy [12], which may improve adherence to surveillance protocols. Capsule endoscopy has become the gold standard for the small bowel, but studies in the oesophagus have yielded mixed results [13]. Wireless capsules have several significant limitations in the oesophagus [14]: the need for a reclined ingestion protocol to increase the imaging period during swallowing from seconds to minutes [13]; difficulty in identifying the capsule location for a given image; and the inability to take biopsies during the procedure.

Tethered capsule endoscopes (Fig. 3.1e) retain the benefits of wireless capsules while addressing several of their limitations in the oesophagus by using a cord to control the capsule's position [15]. The tethered capsule is swallowed by the patient in an upright position, then imaging is performed while it is pulled back up from the stomach. The tethered architecture eliminates the risk of capsule retention and opens up the possibility of capsule re-use, which could lower per-procedure costs [16]. A tethered capsule architecture for Barrett's surveillance is currently in clinical trials [17].

To survey the entire oesophagus during endoscopy, careful articulation of the endoscope is required to bring the entire luminal surface within the 140° forward-facing field of view.

Increased inspection time has been associated with an increased HGD and cancer detection rate [18], but it is unclear whether this is due to additional time inspecting each suspicious site, or additional time articulating the scope to survey the lumen more exhaustively. If the latter is the case, wide angle endoscopes (Fig. 3.1f) [19] with a 330° field of view may allow improved detection rates. Indeed, this has been successfully demonstrated in the colon, decreasing adenoma miss rate from 41 to 7% [20] compared to standard forward-facing devices, although this was not confirmed in a recent randomised study [21]. In the oesophagus, where the surface is smoother with less obscured crevices, wide angle devices may not confer such advantages, but since none have yet been implemented, their potential benefits are unclear. Multiple detectors have also been used to create stereoscopic devices [22], which allow 3D reconstruction of surfaces. It remains unclear whether this innovation will be useful for upper endoscopy, where tissue is relatively smooth.

3.2 The PolyScope Accessory Channel Endoscope

Having reviewed the available device architectures, it was considered that in order to achieve clinical translation within the 4 year timeframe of the work presented in this thesis, an existing accessory channel endoscope device architecture would be most suitable. The PolyScope disposable endoscope (PolyDiagnost, Germany) was chosen

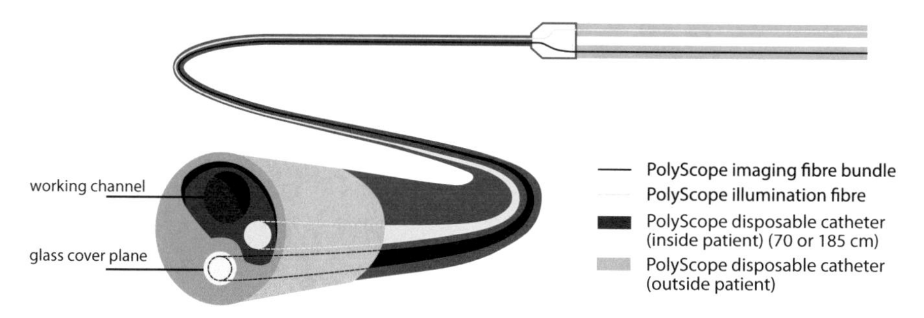

Fig. 3.2 The PolyScope accessory channel endoscope. The PolyScope (PolyDiagnost, Germany) is an accessory channel endoscope CE marked for endoscopic retrograde cholangiopancreatography (ERCP) when passed through the accessory channel of a standard upper gastrointestinal endoscope that passes through the digestive system to the biliary duct. An imaging fibre bundle consisting of 10,000 individual fibrelets is threaded inside a disposable catheter that protects the distal tip of the imaging fibre bundle from direct patient contact with a glass cover plane. The catheter also contains a single fibre light guide for illumination and a 1.2 mm working channel. Catheter lengths of 70 and 185 cm are used in this thesis

(Fig. 3.2). Briefly, an imaging fibre bundle consisting of 10,000 individual fibrelets (PolyDiagnost) is threaded inside a disposable catheter that protects the distal tip of the imaging fibre bundle from direct patient contact with a glass cover plane (PolyDiagnost). The catheter also contains a single fibre light guide for illumination and a 1.2 mm working channel. A range of optical fibre bundles and catheter lengths are available. The PolyScope has several advantages (Fig. 3.3a):

- The PolyScope is CE marked for endoscopic retrograde cholangiopancreatography (ERCP) when passed through the accessory channel of a standard upper gastrointestinal endoscope that passes through the digestive system to the biliary duct. This CE mark means this device is fit for intended purpose, meets legislation relating to safety and can be marketed within the EU, facilitating local approval for safe use in the oesophagus (Translational Characteristic 2).
- Lack of direct patient contact removes the need for sterilisation of the imaging fibre between procedures, prolonging the lifetime of this expensive component of the system. Furthermore, since the only part of the system in contact with the patient is an unmodified disposable catheter, safety approval is further facilitated (Translational Characteristic 2).
- The catheter has a maximum diameter of 3.0 mm, allowing it to be threaded into the 3.7 mm working channel of a commercial standard of care endoscope. This allows; articulation of the device using the standard scope; direct comparison to HD-WLE; and ease of integration into clinical workflow (Translational Characteristic 6).
- The PolyScope allows light to be coupled into and out of the oesophagus, avoiding challenges with miniaturisation of optics, and allowing great flexibility in design of the illumination and detection paths.

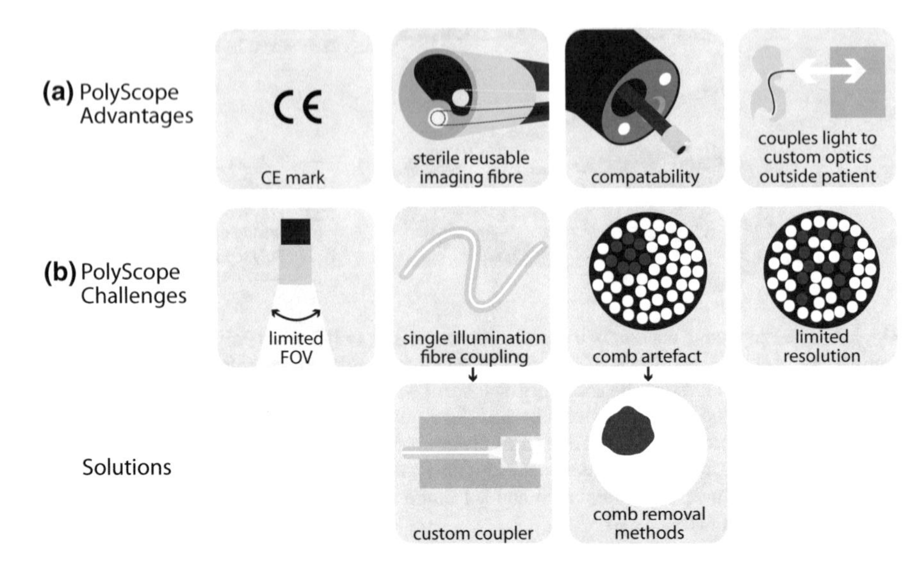

Fig. 3.3 Advantages and challenges of the PolyScope disposable endoscope. **a** The PolyScope system has several advantages: it has a CE mark; the imaging fibre bundle remains sterile and is thus reusable; the PolyScope can be deployed via the working channel of commercial endoscopes; and the PolyScope allows us to couple light to custom optics outside the patient, allowing great flexibility with device design. **b** There are also some challenges. Limited field of view and resolution were accepted as a trade-off for the aforementioned advantages. The single illumination fibre requires careful coupling to ensure adequate light intensity inside the patient, hence the need for a custom coupler. The comb artefact must be removed to achieve images suitable for clinical use by an endoscopist

However, the use of the PolyScope poses several challenges (Fig. 3.3b):

- The field of view (FOV) of the PolyScope is limited. We have two PolyScope fibre bundles, one with a 70° FOV (PD-PS-0093) and another with a 120° FOV (PD-PS-0095), both much less than the typical 140° FOV of standard forward-facing endoscopes. Furthermore, the 120° PolyScope FOV is cropped due to the catheter tip obscuring the edge of the FOV. Despite efforts to remove this, including PolyDiagnost manufacturing a custom 100° scope, this artefact persists, forcing us to accept a limited FOV for our proof of concept device.
- Coupling light into a single 500 μm fibre such that the resulting illumination power at the tip is sufficient for imaging presents a challenge. Initially, a crude butt coupling method interfaced by a PolyDiagnost coupler was used, and illumination power was increased to account for the poor transmission (Sect. 4.4.2). Later, a custom illumination system, including coupling adaptor, was designed and manufactured (Sect. 5.4.2).
- The main challenge in working with the PolyScope arises from the fibre bundle. Despite allowing convenient coupling of light into and out of the oesophagus in a narrow accessory channel device, fiber bundles are associated with low

resolution and a honeycomb like image artefact arising from the opaque cladding between the fibrelets. The resulting image is of low resolution and is displeasing to endoscopists who are accustomed to high resolution HD-WLE images. Since the aim of this work was to develop red flag techniques, which do not rely on high resolution details of the image, but rather on macroscopic reflectance/fluorescence features highlighting suspicious regions, the limited resolution was accepted as a compromise in light of the key advantages previously identified. However, the honeycomb artefact required careful consideration, and will be the subject of the remainder of this chapter.

3.3 Comb Correction Methods

The term 'honeycomb' is generally reserved for the regular hexagonal artefact introduced by a regularly packed fibre bundle, whereas the PolyScope fibre bundle results in an irregular artefact referred to simply as a 'comb' structure [23] (Fig. 3.4). When removing this comb artefact in colour imaging, we must pay careful attention to the impact of the underlying colour filter array (CFA), spectral filters deposited on the pixels of the sensor. Conventional colour cameras have a 2 × 2 Bayer CFA super-pixel of red, green and blue filters (Fig. 3.5a). Spectrally resolved detector arrays (SRDAs) generalise the conventional Bayer CFA with 3 × 3, 4 × 4 or even 5 × 5 CFA super-pixels enabling multispectral imaging (Fig. 3.5b) (Sect. 2.2.3.4) [24, 25]. This introduces a further complication to the image data. A software correction process therefore requires both removal of the comb artefact ("decombing") and separation of the spectral bands ("demosaicking") maintaining both image quality and spectral fidelity [26].

Decombing methods have been extensively explored in the literature and can be readily applied to smooth the appearance of monochrome image data. Winter et al. developed several Fourier-based filtering techniques and compared these to median and Gaussian filters in terms of smoothness (image variance-based) and detail preservation (resolution-based) [23]. They also developed an alternative algorithm to accurately locate and interpolate between fibrelet centres [27, 28], testing a number of interpolation strategies developed in other fields. Standard linear interpolation was found to be most suitable when low processing times are required, as in video-rate imaging [29]. Later work by Lee et al., Regeling et al. and Han et al. further considered different sizes and shapes of Fourier-based filters for decombing, although did not compare these to the other correction strategies [26, 30, 31].

None of these prior studies considered the impact of decombing on signal preservation and as they all used monochrome cameras, they did not address the challenge of combined decombing and demosaicking encountered in colour and more generally, in multispectral imaging. A recent paper by Wang et al. did assess an interpolation strategy in the context of a multispectral fibrescope with an SRDA, but only with respect to the accuracy of the spectral reconstruction and without comparison to other methods [32].

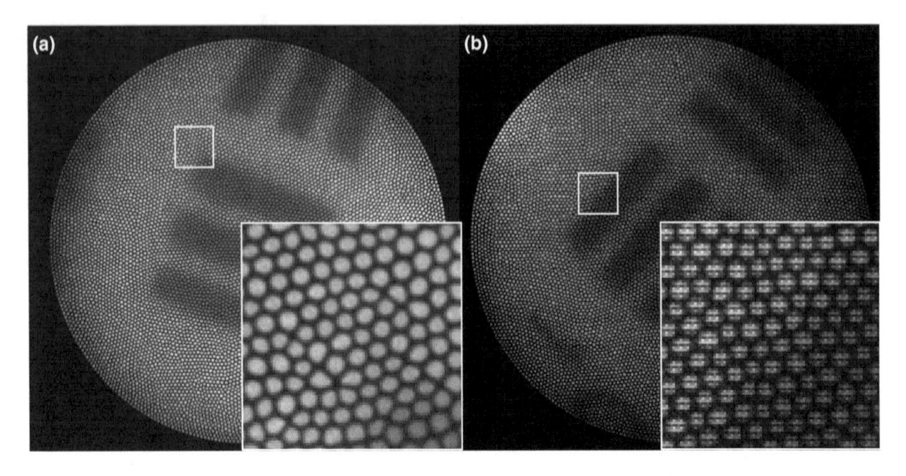

Fig. 3.4 Example image with visible comb artefact and mosaic. **a** Example comb structure from the monochrome fibrescope. Raw image of a USAF test target taken with a monochrome fibrescope (cropped to ~980 × 980). Inset: Zoom of the comb structure showing irregular arrangement and shape of fibrelets. **b** Example comb structure and colour filter array (CFA) mosaic pattern from the multispectral fibrescope. Raw image of a USAF test target taken with a multispectral fibrescope (cropped to ~980 × 950). Inset: Zoom of the comb structure showing the superimposed mosaic due to the pixel-by-pixel transmission variations of the 3 × 3 mosaic of filters deposited pixel-wise on the sensor

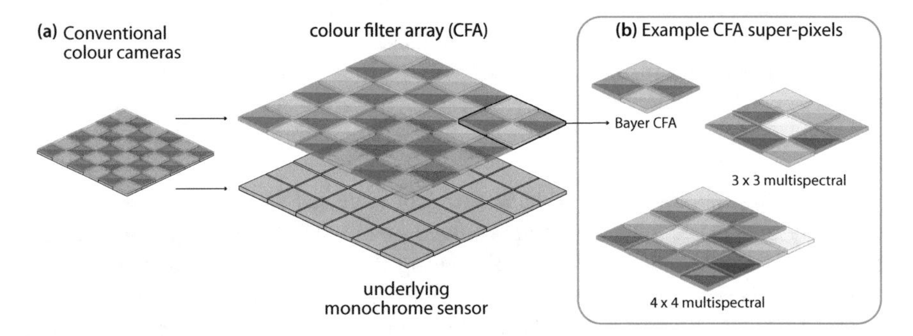

Fig. 3.5 Schematic of spectrally resolved detector arrays (SRDAs) and colour filter arrays (CFAs). **a** Conventional colour imaging is achieved using a colour filter array (CFA), a mosaic of red, green and blue (RGB) colour filters, arranged over a monochrome sensor. **b** The conventional RGB CFA is a Bayer filter. Spectrally resolved detector arrays (SRDAs) generalize the conventional Bayer CFA with 3 × 3, 4 × 4 or even 5 × 5 CFA super-pixels enabling multispectral imaging

To address these limitations, a thorough comparison of decombing techniques combined with demosaicking is presented in this chapter. Performance is assessed with respect to five metrics defined for imaging applications: processing time, resolution, smoothness, signal and accuracy of spectral reconstruction. Filtering methods tested are median, Gaussian and Fourier filtering, compared against interpolation and physical blurring. Initially, performance was evaluated in simulation. Then, an experimental system was designed and used to acquire monochrome and multispectral images, which were used to evaluate performance on real data. Finally, graphs from which the optimum method can be chosen based on the relative importance of each performance metric for a given application were created.

3.3.1 Experimental System

The experimental system was based around the PolyScope (PolyDiagnost, Germany). A narrow band ultra-high power LED (UHP-T-LED-635-EP, Prizmatix, Israel) was coupled into the PolyScope illumination channel using an achromatic doublet lens (AC254-030-A, Thorlabs, Germany) housed inside a custom coupler with a smooth bore for the PolyScope illumination fibre tip (Fig. 3.6a). For reflective samples, external illumination was used to reduce specular reflections (Fig. 3.6b). The detection pathway consisted of an objective lens (NA $= 0.5$, UPLFLN20x, Olympus, Japan) and an achromatic doublet lens ($f = 100$ mm, ACA254-100-A, Thorlabs, Germany), which focused light from the 10,000-fibrelet bundle onto a monochrome CMOS sensor (GS3-U3-41C6M-C, Point Grey, Canada) or a compact SRDA (CMS-V, SILIOS, France) (square pixel sizes 5.5 and 5.3 μm respectively). The SRDA consists of 9 spectral filters (8 narrow bands; average FWHM 30 nm; centre wavelengths 553, 587, 629, 665, 714, 749, 791, 829 nm; 1 broad band; 500–850 nm), deposited as a 3×3 super-pixel across the CMOS sensor. The resulting images using a monochrome camera and multispectral SRDA are shown in Fig. 3.4.

3.3.2 Simulated Images

We simulated monochrome images as outlined in Fig. 3.7 using Matlab® (MathWorks, USA). Briefly, an experimental monochrome comb image was binarised to yield an 'ideal' comb mask; the mask was then magnified by different factors, M, relative to the original in order to investigate the effect of using different fibrelet diameters. 'Ideal' test images were generated as follows: a series of USAF targets to test resolution, an image with uniform intensity to test smoothness and an image with a region of high intensity to test signal. The ideal binary comb masks and the test images were Gaussian blurred to represent imperfections in the fibrelets and imperfections in the test targets respectively. Gaussian noise was added to the ideal test images.

Next, each of the comb masks, $C(x, y)$, was used to mask each test image as follows:

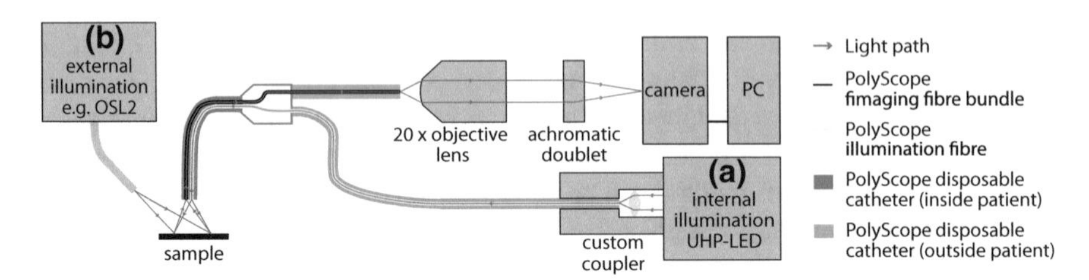

Fig. 3.6 Schematic of the fibrescope. The system is based around the PolyScope accessory channel endoscope (PolyDiagnost, Germany). Illumination is provided either **a** internally, via the PolyScope illumination fibre, with an ultra-high powered LED (UHP-T-LED-635-EP, Prizmatix, Israel) or **b** externally, with a broadband halogen light source (OSL2, Thorlabs, Germany) to reduce specular reflections. Light from the PolyScope 10,000 fibre imaging bundle is focused onto either a monochrome camera (Grasshopper 3, Point Grey, Canada) or an SRDA based multispectral camera (CMS-V, SILIOS, France)

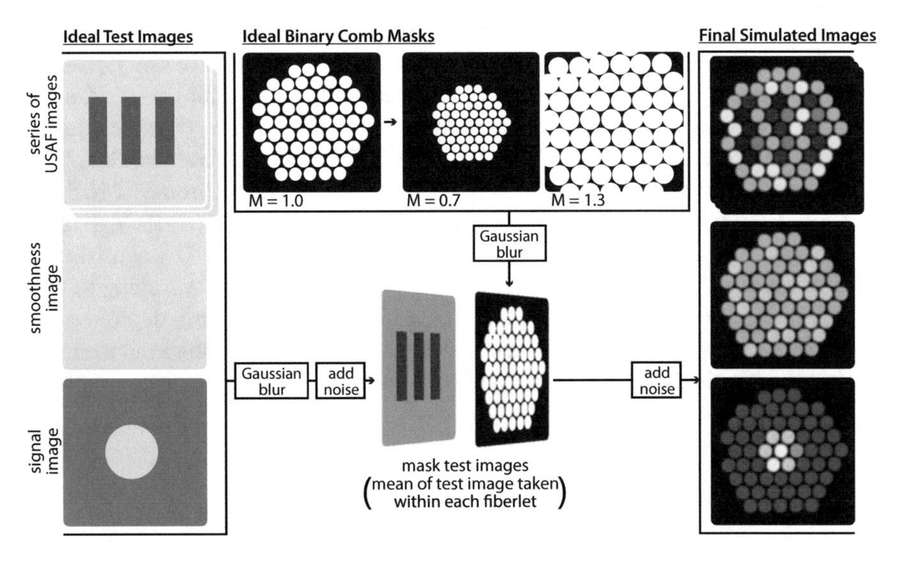

Fig. 3.7 Schematic of the image simulation process. An experimental comb image was binarised to yield an 'ideal' binary comb mask. This was enlarged by a factor, M, compared to the original to create stretched binary comb masks. 'Ideal' test images were generated; a series of USAF targets to test resolution, an image with uniform intensity to test smoothness and an image with a region of high intensity to test signal. The ideal binary comb masks and the test images were blurred to represent imperfections in the fibrelets and imperfections in the test targets respectively. Noise was added to the ideal test images. Next, each of the comb masks was used to mask each test image as follows. The comb mask was split into individual fibrelet masks and for each fibrelet the mean of the test image in the region where the intensity of the fibrelet mask was >0.1 was taken, then multiplied by the fibrelet mask, and added to the final image. Noise was added to reach the final simulated images

(i) The comb mask, $C(x, y)$, was split into individual fibrelet masks, $C_i(x, y)$.
(ii) Each individual fibrelet mask was binarised to generate binarised individual comb masks:

$$C_i^B(x, y) = \begin{cases} 1, & C_i(x, y) > 0.1 \\ 0, & C_i(x, y) \leq 0.1 \end{cases} \tag{3.1}$$

(iii) In the region defined by $C_i^B(x, y)$, the mean of the test image, $T(x, y)$, was taken:

$$M_i = \sum_{x,y} T(x, y) C_i^B(x, y) / \sum_{x,y} C_i^B(x, y) \tag{3.2}$$

(iv) This was multiplied by the individual fibrelet masks, $C_i(x, y)$, to generate the simulated image, I^{sim}:

$$I^{sim}(x, y) = \sum_i C_i(x, y) M_i \tag{3.3}$$

Finally, Gaussian noise was added to reach the final simulated images.

3.3.3 *Monochrome Image Comb Correction*

All image analysis was performed in Matlab® (MathWorks, USA). Five different decombing methods were applied directly to simulated and experimentally captured monochrome images. For all methods, the amount of filtering is defined by a dimensionless characteristic filter size r, which is varied. *Gaussian blur* was achieved using the Matlab function 'imgaussfilt' to filter the images with a 2D Gaussian smoothing kernel with a standard deviation of r ($1 < r < 50$ pixels). *Median filtering* was achieved using the Matlab function 'medfilt2' to filter images by taking the median value of each $2r$-by-$2r$-pixel region ($1 < r < 50$ pixels). *Fourier filtering* makes use of the Matlab fast Fourier transform function, 'fft2'. The Fourier transform image is cropped to remove any frequencies above f_o, the cut-off frequency, and inverse Fourier transformed, using 'ifft2', to obtain the corrected image. The cut-off frequency was defined as:

$$f_o = \frac{1}{2r} \tag{3.4}$$

where r is the characteristic filter size input ($1 < r < 50$ pixels) and corresponds to the smallest resolvable feature size in the inverse Fourier transformed image according to Nyquist's theorem.

Interpolation relies on knowing the location of each individual fibrelet in the image, for which we used a centre finding algorithm based on the work of Elter et al. [27]:

(i) Acquire a bright field calibration image, I, of a white reflectance target.
(ii) Candidate points are selected based on their brightness in relation to their local neighbourhood. Around each pixel (x, y) neighbourhood is defined as:

$$N = I(x - d : x + d, y - d : y + d) \tag{3.5}$$

where the size of the neighbourhood, D, can be defined as:

$$D = 2d + 1 \tag{3.6}$$

where d is specified by the user such that D is roughly the diameter of a single fibrelet.
Given a minimal intensity difference I_{min}, which is specified by the user as the expected minimum intensity difference between fibrelets and cladding, a pixel (x, y) is considered as a candidate centre point if:

$$\max(N) - \min(N) > I_{min} \tag{3.7}$$

(iii) For each candidate centre point, (x_c, y_c), a score, s_c, is calculated indicating how well a 2D quasi-Gaussian surface fits the neighbourhood around this point:

$$s_c = \sum_N (G(x, y) - I(x, y))^2 \tag{3.8}$$

where G is a defined as:

$$G = He^{-l^4/2D^2} \tag{3.9}$$

with l the distance from the candidate point and $H = I(x_c, y_c)$.
(iv) Order candidate centre points (x_c, y_c) by ascending score s_c.
(v) Starting from the highest ranked centre (lowest score), sequentially place each candidate fibre centre (x_c, y_c) onto a centre map *if and only if* the candidate centre is a minimum distance of one fibre diameter from all centres (x_m, y_m) already in the map:

$$\sqrt{(x_c - x_m)^2 + (y_c - y_m)^2} > D, \quad \forall m \tag{3.10}$$

(vi) Fibre centres are added for as long as this criterion is satisfied until all candidates have been added to the map or rejected.

Decombing is then achieved using bilinear interpolation of the pixel values recorded at the fibre centres, $I(x_m, y_m)$.

As a reference gold standard, we also performed physical blurring of our image by experimentally defocusing the image of the fiberscope face. We displaced the fiberscope along the optical axis by 5 μm in 0.5 μm steps in both directions and recorded images. In order to plot physical blurring alongside the other metrics, the dimensionless characteristic filter size, r, is arbitrarily defined as:

$$r = \left| \frac{\text{displacement}}{0.5\mu\text{m}} \right| \tag{3.11}$$

3.3.4 Multispectral Image Comb Correction

Multispectral images consist of a mosaic of spectral information due to the filter deposition pattern of the CFA. In a process known as demosaicking (Fig. 3.8a), the final colour image is reconstructed by splitting the raw camera output into 9 incomplete mosaic pattern images that are subsequently interpolated. Demosaicking must occur prior to decombing with Gaussian, median and Fourier filtering, as mixing information from adjacent pixels on the raw image would corrupt the recorded spectral information. Gaussian, median and Fourier filtering were applied in the same way as for monochrome images (Sect. 3.3.3). When physical blurring was applied to the image at the point of capture, light spreads from each fibrelet, decombing the image prior to its passage through the CFA, so following demosaicking, no further filtering takes place. Interpolation between fibrelet centres must occur in parallel to demosaicking (Fig. 3.8b). This was implemented as follows:

(i) Acquire a bright field calibration image, $I(x, y)$, with broadband illumination to ensure a signal is recorded in all spectral bands.
(ii) Apply a Gaussian blur to the bright field calibration image in order to smooth out the mosaic pattern due to the CFA.
(iii) Find the centres of the fibrelets, (x_m, y_m), using steps ii–vi of the centre finding algorithm outlined in Sect. 3.3.3.
(iv) At each fibrelet centre point (x_m, y_m), we need to know the image intensity in each spectral band k, $I^k(x_m, y_m)$. The centre point corresponds to a pixel with one spectral filter, giving the intensity in one of the spectral bands. For the other spectral bands, the intensity at the centre point is assumed to be the same as the intensity at the nearest neighbour with the correct spectral filter. The filter deposition pattern on the sensor, $P(x, y)$, is known:

$$P(x, y) = \begin{cases} 1 \\ 2 \\ \vdots \end{cases} \tag{3.12}$$

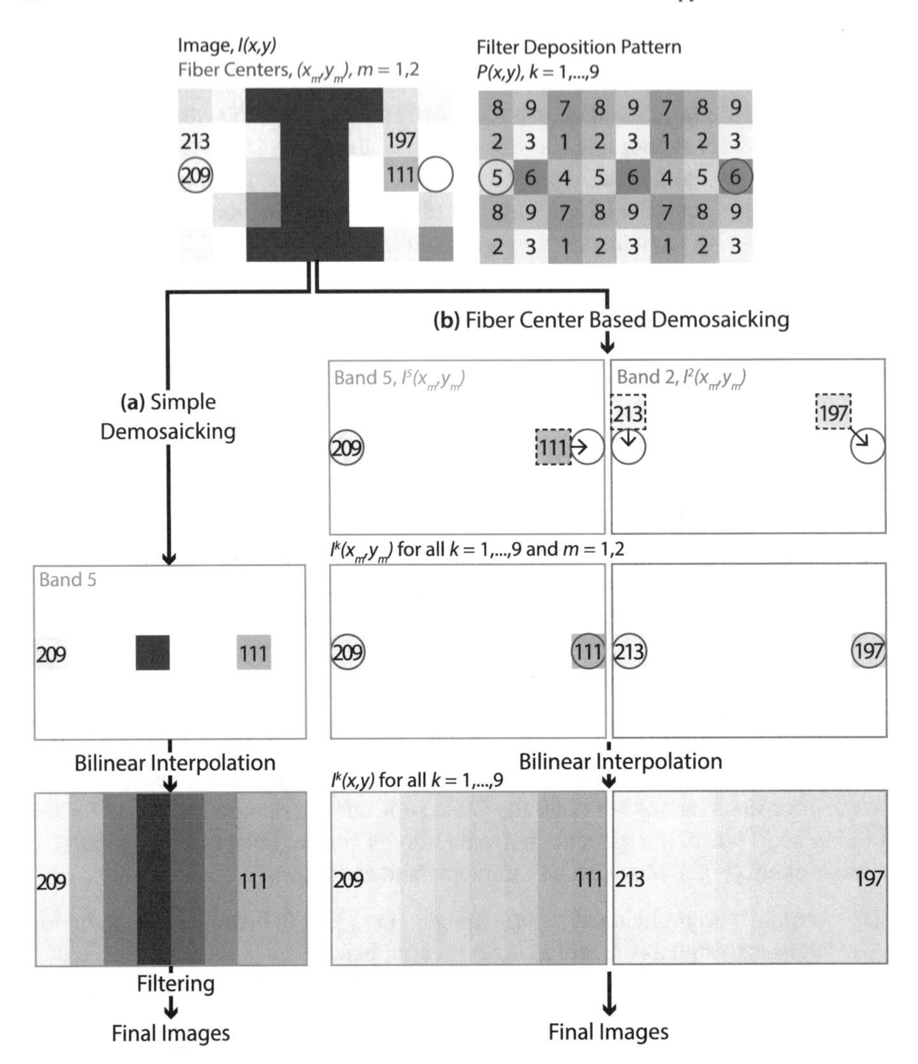

Fig. 3.8 Schematic of multispectral demosaicking and comb removal algorithms. An example image is shown with two fibrelet centres. The corresponding filter deposition pattern for a 9-band colour filter array is shown to its right. **a** Simple demosaicking according to the filter deposition pattern is shown for band 5. Following this, bilinear interpolation is used to fill out the image. Finally, filtering of each image occurs, with either a Gaussian, median, or Fourier filter applied, or no filter in the case where physical blurring was applied to the image at the point of capture. **b** Fibre centre based demosaicking is shown for bands 5 and 2. Briefly, at each fibre centre location, the intensity in each band is determined by taking the intensity at the nearest neighbouring pixel with the corresponding band filter. This is followed by interpolation between fibre centres, which can be implemented with a single look up table, since in every band and every image we interpolate between the same centre points

So, the image intensity, $I^k(x_m, y_\mathrm{m})$, can be defined as:

$$I^k(x_m, y_\mathrm{m}) = I(x_{NN}, y_{NN}) \tag{3.13}$$

where (x_{NN}, y_{NN}) is the nearest neighbour pixel with the spectral filter k:

$$P(x_{NN}, y_{NN}) = k \tag{3.14}$$

This process should result in image data for all spectral bands and at all centre locations:

$$I^k(x_m, y_m) \quad \forall k, m \tag{3.15}$$

This is generalisable to any size mosaic as long as the ratio, T, between the superpixel size and the fibrelet size on the sensor, obeys the criterion:

$$T = \frac{L}{D} < 1 \tag{3.16}$$

where L is the size of a super pixel and D is the diameter of the fibrelets on the sensor in pixels. Though not used here, even super pixels (e.g. 2×2, 4×4) can result in some image points having two equally distant nearest neighbours in some spectral bands, so an appropriate randomized selection or average of these would need to be taken.

Finally, using bilinear interpolation of the pixel values at fibre centres within each spectral band, $I^k(x_m, y_m)$, a comb free image for each spectral band, $I^k(x, y)$, is constructed.

3.3.5 Performance Metrics

In order to determine the performance of comb removal, 4 performance metrics for monochrome imaging and 5 performance metrics for multispectral imaging were assessed.

3.3.5.1 Resolution

Resolution was determined by capturing images of a 1951 USAF resolution test target (#53-714, Edmund Optics, USA) externally illuminated with a broadband halogen light source (OSL2, Thorlabs, Germany) to reduce specular reflections. The Michelson contrast, C^M, was calculated for each element:

$$C^M = \frac{I_{max} - I_{min}}{I_{max} + I_{min}} \tag{3.17}$$

where I_{max} and I_{min} are the maximum and minimum image intensities in each USAF element respectively.

The resolution, R, was determined as the line width when Michelson contrast dropped below 5% by fitting a smoothing spline to the Michelson contrast versus the line spacing. A contrast threshold of 1% has previously been reported to be applicable across a wide range of targets and conditions [33], but a contrast threshold of 5% was chosen to avoid effects arising from noise at very low contrast. A condition requiring data points with contrast >1% at >3 distinct line spacings was also implemented to ensure there were a reasonable number of non-noise data points to fit a spline. If this condition was not met, a spline was not fitted and the resolution was not defined. For multispectral imaging, the resolution is taken as the average of the resolution determined separately for each of the 9 spectral bands. The resolution score [23], S_{res}, was defined as:

$$S_{res} = 1 - \left(\frac{R - R_{min}}{R_{max} - R_{min}} \right) \tag{3.18}$$

where R_{max} and R_{min} are the maximum and minimum resolutions calculated across all correction methods, defined such that scores of 1 and 0 represent the best and worst resolution achieved respectively.

3.3.5.2 Smoothness

For monochrome imaging, smoothness was determined by using images of white areas of a 1951 USAF resolution test target (#53-714, Edmund Optics, USA) illuminated externally with a broadband halogen light source (OSL2, Thorlabs, Germany) to reduce specular reflections. For multispectral imaging, smoothness was determined using images of a white reflecting target (paper) illuminated with a narrow band LED (UHP-T-LED-635-EP, Prizmatix, Israel) coupled to the PolyScope illumination fibre.

The spatial standard deviation of the supposedly uniform image was calculated. The average of this across 18 images was taken for multispectral imaging (9 spectral bands from each of the two light sources). The smoothness score [23], S_{smooth}, was defined as:

$$S_{smooth} = 1 - \left(\frac{\sigma - \sigma_{min}}{\sigma_{max} - \sigma_{min}} \right) \tag{3.19}$$

where σ_{max} and σ_{min} are the maximum and minimum standard deviations calculated across all correction methods, such that scores of 1 and 0 represent the best and worst smoothness achieved respectively.

3.3.5.3 Signal

It is crucial that decombing and demosaicking preserve regions of high pixel intensity, as detection of these high signal regions is central to techniques such as optical molecular imaging (OMI), autofluorescence imaging and reflectance imaging. Fluorescence emission was chosen as the source used to quantify signal here, as it is relevant to our work on OMI (Chap. 4), but ultimately, the physical source of the regions of high pixel intensity should not alter the results.

Fluorescence signals were acquired by capturing images of a 30 μL solution of 1 mg/mL of the fluorescent dye AF647 (Thermo Fisher Scientific, USA) dissolved in phosphate buffered saline (PBS) in a well plate (μ-Slide 18 Well Flat, ibidi GmbH, Germany) using illumination from a 635 nm LED (UHP-T-LED-635-EP, Prizmatix, Israel) and a long pass emission filter in front of the camera (ET700/75m, Chroma, USA). The mean pixel intensity, S, was calculated in a region of interest (ROI) drawn manually inside the well on the image. For multispectral imaging, the mean pixel intensity was extracted from those bands that overlap with the emission spectrum of AF647 (narrow bands: 665, 714 nm, FWHM, 27, 26 nm; broadband: 500–850 nm). The signal score, S_{signal}, was defined as:

$$S_{signal} = \frac{S - S_{min}}{S_{max} - S_{min}} \tag{3.20}$$

where S_{max} and S_{min} are the maximum and minimum signals calculated across all correction methods, such that scores of 1 and 0 represent the best and worst signal achieved respectively.

3.3.5.4 Speed

Speed was determined by measuring the total computation time to correct 10 images in series on a MacBook Pro (Processor 2.4 GHz Intel Core i5, Memory 8 GB 1600 MHz DDR3) and then dividing by 10 to calculate the time per frame.

3.3.5.5 Accuracy of Spectral Reconstruction (Multispectral Performance Metric Only)

For multispectral imaging, it is crucial that the information from different pixels on the SRDA is not mixed by the comb correction process. In order to assess the performance of each correction method, a score to represent the accuracy of spectral reconstruction (ASR) was defined. To extract this score, the following steps were performed:

(i) Using the multispectral endoscope, an image of a white reflecting target (paper), illuminated by a narrowband source (UHP-T-LED-635-EP, Prizmatix, Israel) coupled to the PolyScope illumination fibre, was captured.
(ii) This image was demosaicked and decombed as outlined in Sect. 3.3.4 to provide a multispectral cube of data: $I^k,\ k = 1-9$ where k indicates the spectral band.
(iii) The 'ground truth' spectral properties of the image data were determined. The spectrum, $G(\lambda)$, of the target was captured using a reference spectrometer (AvaSpec-ULS2048, Avantes, Netherlands).
(iv) The 'ground truth' spectrum was multiplied by the response of the endoscope in each spectral band k, $R^k(\lambda)$, to predict the 'ground truth' recorded spectrum:

$$E^k = \sum_{\lambda} G(\lambda) R^k(\lambda), \quad k = 1-9 \tag{3.21}$$

(v) The normalised (to AUC = 1) average (over all pixels in the image) spectrum collected with the endoscope was compared to the predicted 'ground truth' spectrum and the mean squared difference was determined by:

$$Q = \sum_{k=1}^{9} \left(\overline{\left(\frac{\sum_{x,y} I^k(x,y)}{\sum_{x,y} 1} \right)} - \overline{E^k} \right)^2 \tag{3.22}$$

where the bar represents normalisation of spectra to AUC = 1.
(vi) The accuracy of spectral reconstruction score, S_{ASR}, is defined as:

$$S_{ASR} = 1 - \left(\frac{Q - Q_{min}}{Q_{max} - Q_{min}} \right) \tag{3.23}$$

where Q_{max} and Q_{min} are the maximum and minimum mean squared differences calculated across all correction methods, such that scores of 1 and 0 represent the best and worst accuracy of spectral reconstruction achieved respectively.

By spatially (pixel-by-pixel) averaging the spectra prior to normalisation and calculation of the mean square difference, the influence of non-smoothness of the images is reduced, since this effect is already accounted for in the smoothness metric.

3.3.6 Overall Performance

Since there are trade-offs between the performances of the metrics, the overall performance of a particular correction method depends on which of the metrics are prioritised in a given application. To account for this, an overall performance score, *OP*, was constructed:

$$OP = w_{res}S_{res} + w_{smooth}S_{smooth} + w_{signal}S_{signal} + w_{ASR}S_{ASR} \tag{3.24}$$

with adjustable application-dependent weightings, *w*, to emphasize a priority metric, such that:

$$w_{res} + w_{smooth} + w_{signal} + w_{\mathrm{ASR}} = 1 \tag{3.25}$$

Speed was not included in *OP* since it is possible to independently optimize speed by improving hardware and parallelizing software. For monochrome imaging $w_{ASR} = 0$. Since weightings are application-dependent, the *OP* for all weightings was calculated, and displayed these using a colour map such that the reader may visually select the optimum correction method.

3.4 Comb Correction Results

3.4.1 *Performance of Simulated Monochrome Image Corrections*

The scores for resolution, smoothness and signal as a function of the characteristic filter size for 4 decombing methods tested in simulated monochrome images were calculated for three different sized comb structures, M = 0.7, M = 1.0 and M = 1.3 (Fig. 3.9). These structure sizes were chosen to represent a range from 7500–30,000 fibrelets per bundle, magnified to fill a similar ~1000 × 1000 pixel sensor. For physical blurring, the correction occurs in hardware at the point of imaging. The current approach does not replicate the full imaging process, so accurate simulation of physical blurring was not possible. Speed was not tested with simulated mages, since the size of the comb structure makes no difference to computation time.

These simulations revealed that the size of the fibrelet has little effect on the overall trends in performance of the different methods. This is confirmed in Fig. 3.10, which shows the overall performance for the different fibrelet sizes. The general shape of the optimal performance space remains constant, with the only change being a preference for median filtering in a tiny region of the performance space (around $w_{smooth} = 0.85$, $w_{signal} = 0.15$) for the smaller fibrelet dimensions (M = 0.7). This is because median filtering performs better when the filling factor, the fraction of the image filled with data (fibrelet) versus artefact (cladding), is larger, as is the case for the smaller fibrelet size. This is also observed in Fig. 3.9, where the M = 0.7 median data follows a different trend to the M = 1.0 and M = 1.3 data, giving a higher signal score.

Since fibre diameter made little difference to the overall layout of performance space, we continued our investigation using experimental data from our endoscope, which produced images corresponding to M = 1.0.

3.4.2 Performance of Experimentally Captured Monochrome Image Corrections

The scores for resolution, smoothness, signal and speed as a function of the characteristic filter size for the 5 decombing methods tested in monochrome imaging enable direct comparisons to be made regarding the individual performance of the methods (Fig. 3.11). Example images are shown in Fig. 3.12.

Similar trends are observed for performance of corrections in experimental image data (Fig. 3.11) and in simulated image data (Figs. 3.9 and 3.10). The main difference between simulation and experiment is that interpolation appears to achieve an enhanced resolution score in experimental data compared to simulated data, suggesting the noise component of simulated data may have been slightly overestimated.

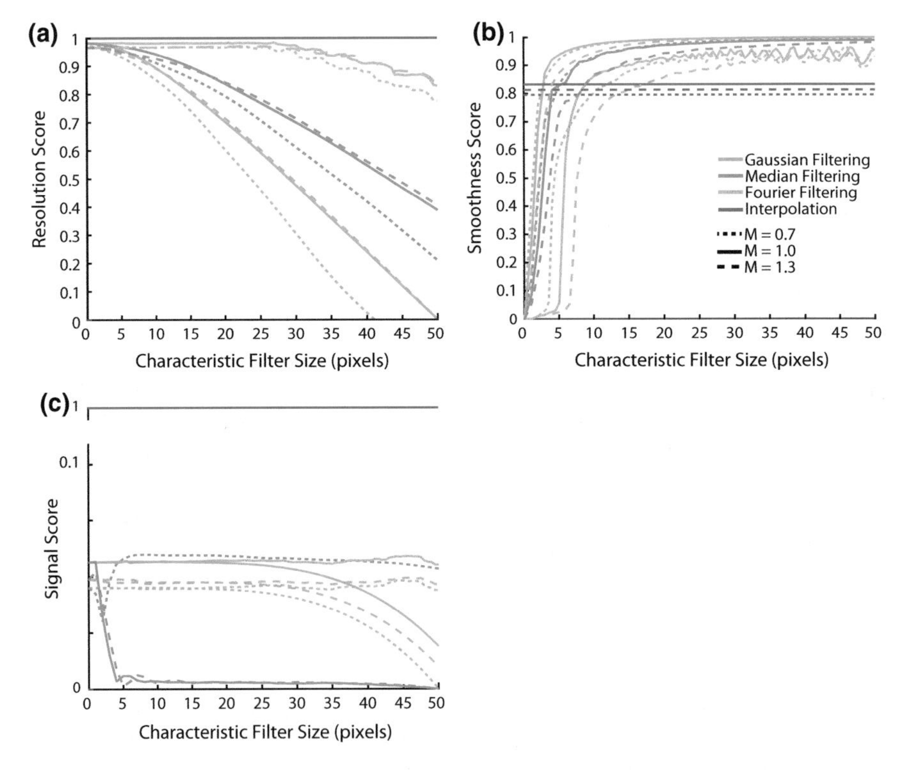

Fig. 3.9 Performance scores for 4 correction methods applied to simulated monochrome images for M = 0.7, M = 1.0 and M = 1.3. **a** Resolution score. R_{max} = 82, 102, 103 pixels (648, 803, 809 μm) and R_{min} = 21.4, 22.4, 23.5 pixels (169, 177, 185 μm) for M = 0.7, 1.0, 1.3 respectively. **b** Smoothness score. σ_{max} = 20.3, 19.9, 20.1 and σ_{min} = 0.638, 0.388, 0.346 for M = 0.7, 1.0, 1.3 respectively. **c** Signal score. S_{max} = 151, 156, 156 and S_{min} = 118, 117, 118 for M = 0.7, 1.0, 1.3 respectively. Dotted line M = 0.7, solid line M = 1.0, dashed line M = 1.3. Since interpolation between irregularly spaced points is a complex spatially variant filter, the characteristic filter size is not well defined, so the score for interpolation is represented as horizontal line in each graph

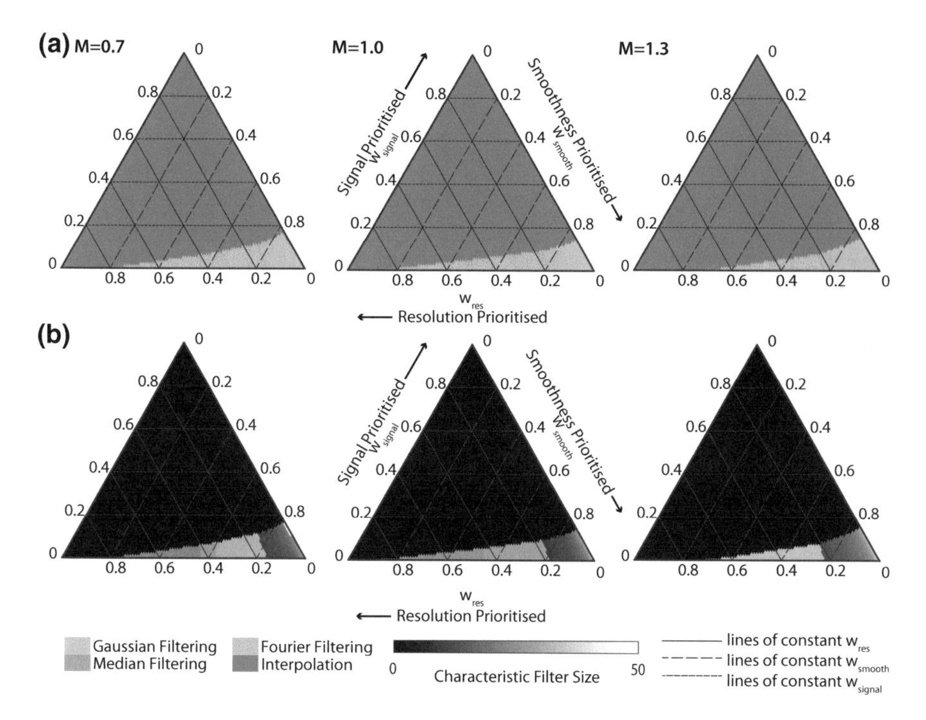

Fig. 3.10 Optimum correction method based on overall performance score for simulated images with M = 0.7, M = 1.0 and M = 1.3. **a** The correction method that gives highest overall performance (OP). **b** The characteristic filter size used with this correction method to achieve the highest OP. Since interpolation between irregularly spaced points is a complex spatially variant filter, the characteristic filter size is not well defined, so it is represented as zero in the graphs. Speed is not included in the OP since it is possible to optimise speed independently

Overestimation of the noise component would disproportionately degrade the performance of interpolation relative to the other correction methods; it is more susceptible to noise as it relies on data from a single pixel per fibrelet.

The results from the monochrome data corrections are summarised in Fig. 3.11. For preservation of resolution and signal, interpolation clearly provides the optimal solution. If image smoothness is our only priority, Gaussian filtering is preferred. Otherwise, Fourier-filtering provides a compromise between smoothness and resolution.

3.4.3 Performance of Captured Multispectral Image Corrections

The scores for resolution, smoothness, signal, ASR and speed as a function of the characteristic filter size for the 5 decombing and demosaicking methods tested in

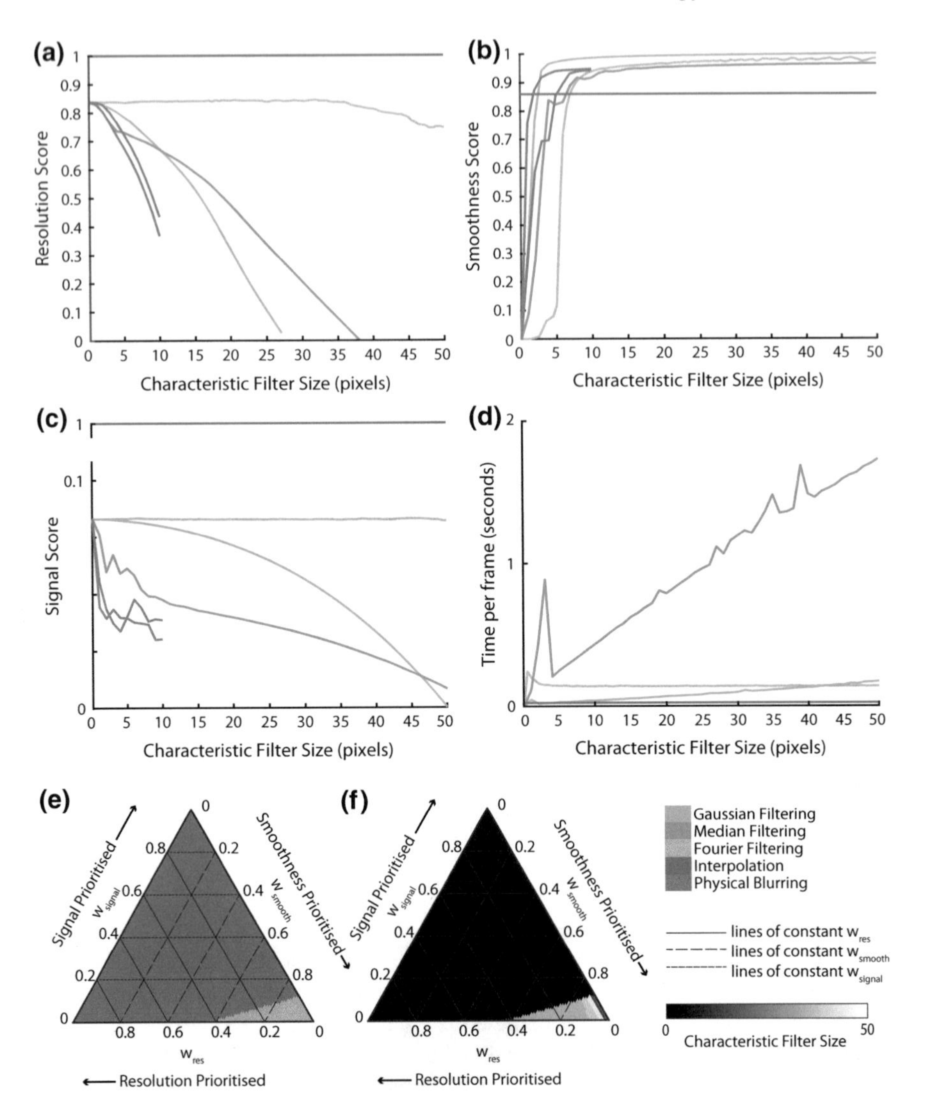

Fig. 3.11 Performance scores for 5 correction methods applied to monochrome images. **a** Resolution score. $R_{max} = 500$ µm and $R_{min} = 228$ µm. **b** Smoothness score. $\sigma_{max} = 32.2$ and $\sigma_{min} = 3.85$. **c** Signal score. $S_{max} = 157$ and $S_{min} = 114$. **d** Time to correct each frame. Since interpolation between irregularly spaced points is a complex spatially variant filter, the characteristic filter size is not well defined, so the score for interpolation is represented as horizontal line in each graph. **e** Optimum correction method based on highest overall performance score OP for weightings w_{res}, w_{smooth} and w_{signal}. **f** The characteristic filter size used with this correction method to achieve the highest overall performance (OP). Since interpolation between irregularly spaced points is a complex spatially variant filter, the characteristic filter size is not well defined, so the score for interpolation is represented as zero the graphs. Speed is not included in the OP since it is possible to optimize speed independently

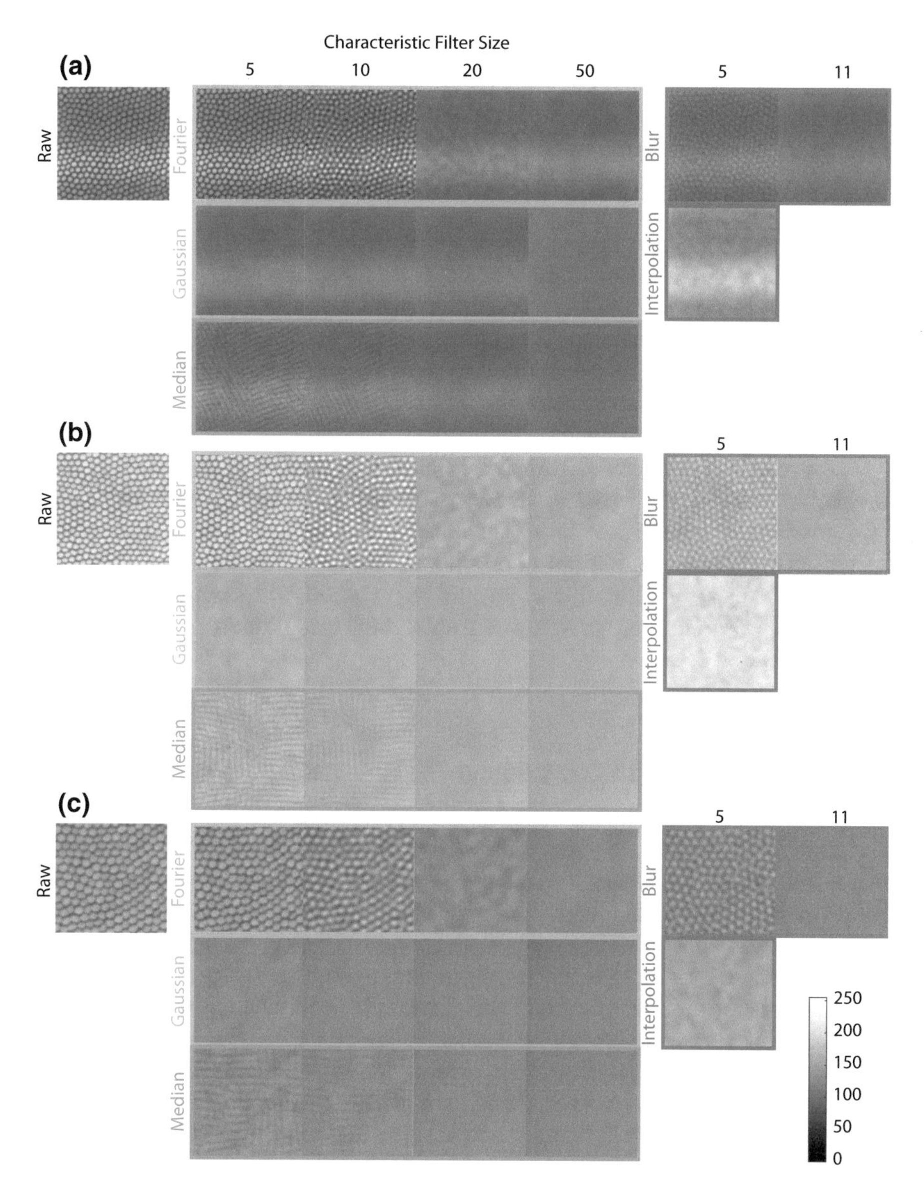

Fig. 3.12 Example corrected monochrome images. **a** Region of an image of a USAF test chart element used to determine resolution. **b** Image of white reflecting target (paper) used to determine smoothness. **c** Image of AF647 used to determine signal

multispectral imaging are shown in Fig. 3.13. The ASR score is highest for interpolation, which is expected since interpolation removes the erroneous pixels from the cladding region. The other methods yield reduced ASR scores since they mix together the 'true' spectra from fibrelet centres and the 'erroneous' spectra from the cladding, which is particularly evident in median filtering, where the sparsity of 'true' pixels, from the compound sampling of mosaic and comb can result in median filters removing 'true' pixels over the more common 'erroneous' pixels, and in physical blurring, where the 'true' and 'erroneous' spectra are mixed before detection.

The computation time in multispectral corrections is increased by around one order of magnitude, as expected since there are nine images to correct rather than one. The computation time of physical blurring represents demosaicking alone. If direct demosaicking were performed, resulting in 9 images each with a 9-fold reduction in total image pixels, the computation time would be significantly reduced.

The overall performance is shown in Fig. 3.14. In the context of multispectral imaging, it was found that interpolation frequently provides the optimal solution for correction as was the case for monochrome imaging. Only when smoothness is prioritised does interpolation fail to provide the best solution. In this special case, Gaussian, median or Fourier filtering would be preferred, with Fourier filtering providing a better solution preservation of resolution is also desired. Example images are shown in Fig. 3.15.

3.5 Discussion and Conclusions

The clinically approved PolyScope disposable endoscope allows light to be delivered to, and collect light from, the oesophagus, enabling the development and in vivo evaluation of novel optical imaging techniques with the capacity for swift translation. However, the use of a fibre bundle introduces a comb structure into the images, which must be removed by image decombing algorithms to improve image quality to a level that is satisfactory for the end user. This process is further complicated in SRDA-based multispectral imaging, where it must be performed alongside demosaicking. In order to address this problem, the performance of commonly used methods of decombing was evaluated and compared in monochrome imaging and SRDA-based multispectral imaging. By defining scores relating to resolution, smoothness, signal, speed and accuracy of spectral reconstruction, each of these correction methods was evaluated within the parameter space of relevance for fiberscope imaging. Through data acquired with simulation and imaging, it was shown that interpolation provides the optimal solution in most cases for both monochrome and multispectral imaging. Only when image smoothness is highly prioritised, is Gaussian filtering preferred, while Fourier-filtering can be used in cases requiring a compromise between smoothness and resolution.

The optimal choice of correction method is application-dependent, hence rather than providing absolute recommendations, overall performance scores composed of weighted sums of performance metrics were objectively prepared. Weightings can

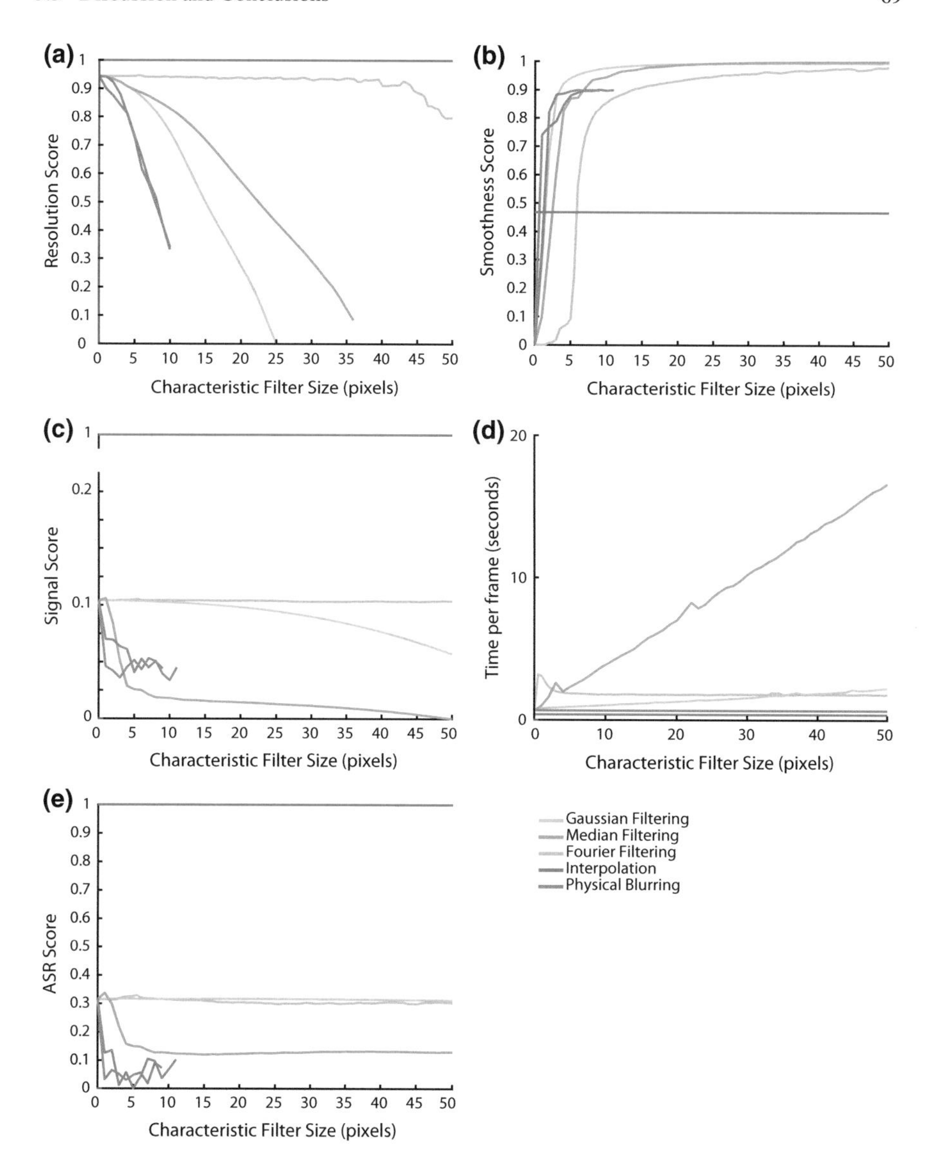

Fig. 3.13 Performance scores for 5 correction methods applied to multispectral images for multispectral images. **a** Resolution score. $R_{max} = 440\ \mu m$ and $R_{min} = 260\ \mu m$. **b** Smoothness score. $\sigma_{max} = 13$ and $\sigma_{min} = 0.73$. **c** Signal score. $S_{max} = 28.1$ and $S_{min} = 10.6$. **d** Speed. **e** Accuracy of spectral reconstruction. $Q_{max} = 0.0138$ and $Q_{min} = 0.0060$. Since interpolation between irregularly spaced points is a complex spatially variant filter, the characteristic filter size is not well defined, so the score for interpolation is represented as horizontal line in each graph

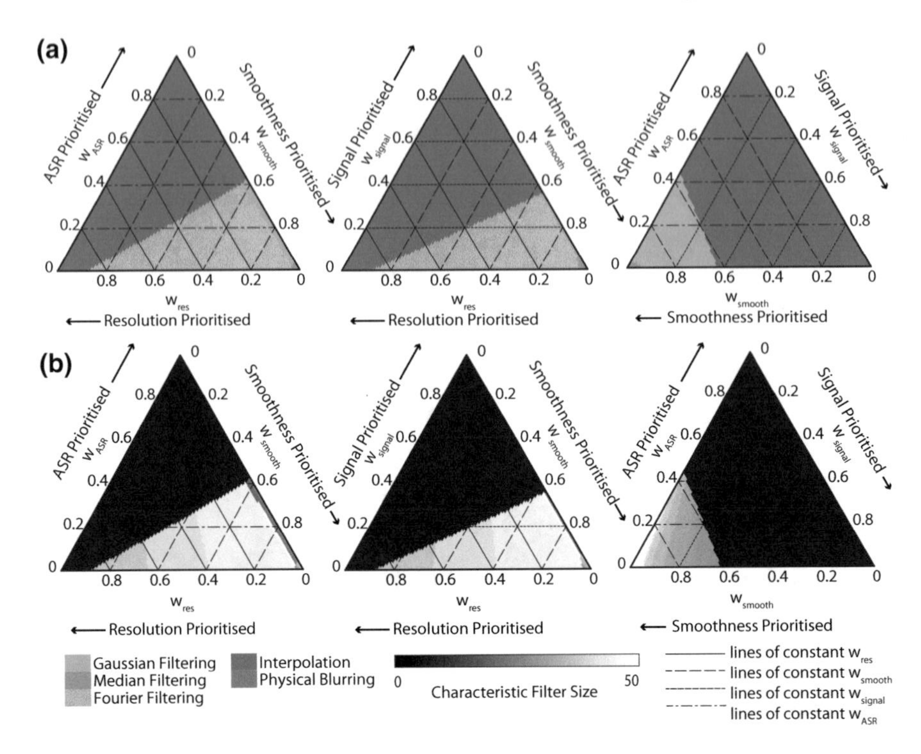

Fig. 3.14 Optimum correction method based on overall performance score. **a** The correction method that gives highest overall performance (OP). **b** The characteristic filter size used with this correction method to achieve the highest OP. For each ternary plot, the weighting not shown on the axes is set to zero. Since interpolation between irregularly spaced points is a complex spatially variant filter, the characteristic filter size is not well defined, so the score for interpolation is represented as zero in the graphs. Speed is not included in the OP since it is possible to optimise speed independently. The trade-off between resolution, signal and ASR, is not shown since in all combinations interpolation gives the best overall performance

therefore be assigned depending on the imaging priorities in a given application, which will vary significantly. To give some examples, visual inspection in real time puts speed as a high priority while inspection of detailed surface features, such as mucosal patterns or vasculature, will put resolution as a high priority. For autofluorescence imaging or fluorescence molecular imaging, where signal is relatively limited due to low abundance of fluorescence molecules in vivo, signal should be maximised to facilitate detection of these fluorescence markers. In cases requiring supervised classification of spectral signatures, such as in evaluation of oxy- and deoxy-haemoglobin concentrations during multispectral imaging, the accuracy of the spectral reconstruction is vital.

While the results presented here enable comparison of the performance of these different correction methods across a range of applications, there remain some limitations to this study. Firstly, only the case where the fibrelet image size on the sensor

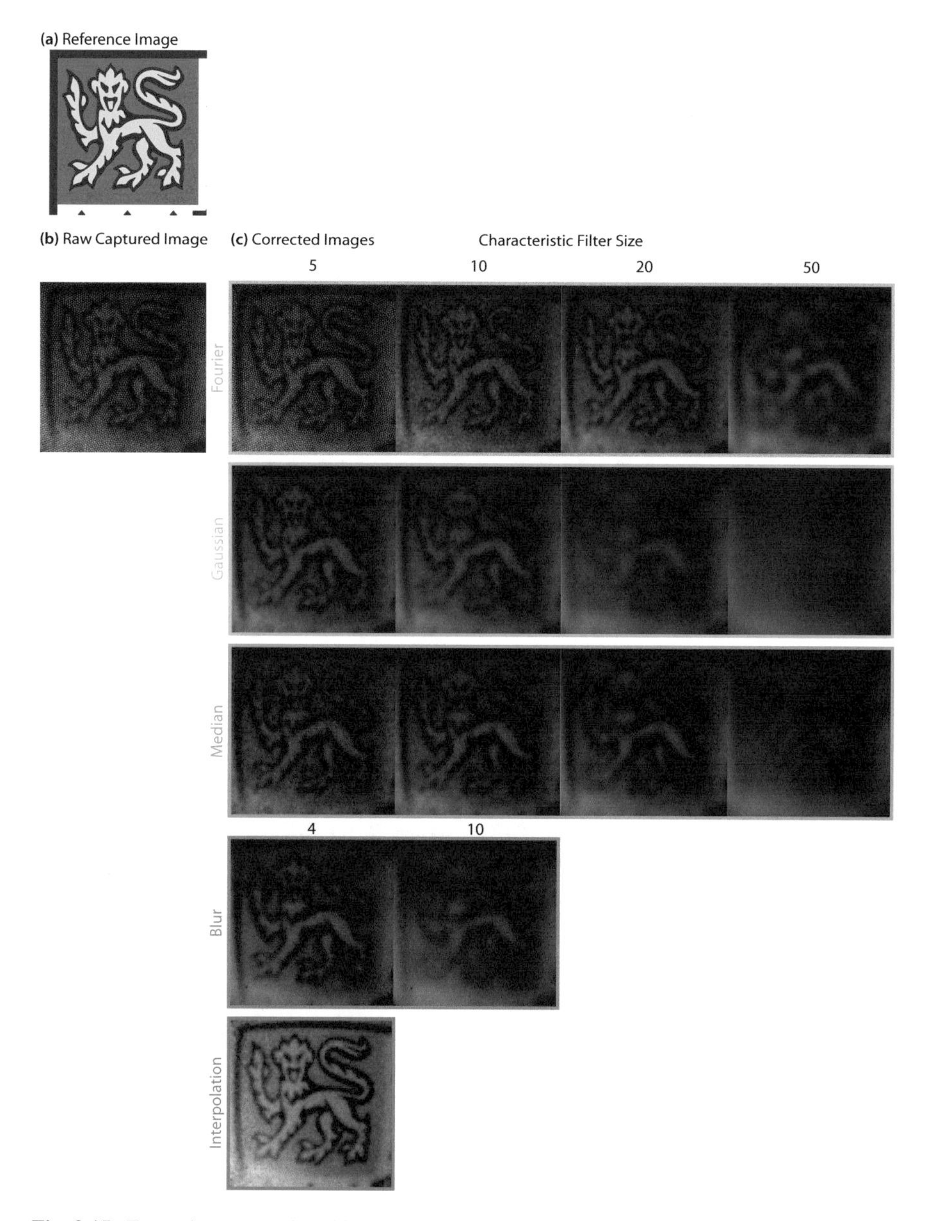

Fig. 3.15 Example corrected multispectral images. **a** A section of the University of Cambridge crest. This was printed and imaged with the multispectral endoscope. **b** The image generated by demosaicing the raw captured image and assigning the narrow bands centred at 629, 587 and 553 nm to RGB channels respectively. **c** Corrected images

is greater than the size of a super-pixel ($T < 1$) was considered. In practical applications, appropriate lenses could be used to magnify the image to ensure that this constraint is fulfilled. A consequence of this constraint is that demosaicking should not affect resolution. Indeed, the monochrome sensor and multispectral sensors had maximum resolutions of 228 ± 18 and 240 ± 20 μm respectively (errors determined from errors in a linear fit to Michelson contrast data). Secondly, the magnification approach that was used in simulation does not quite recapitulate all possible combs, since it magnifies both the cores and the cladding, whereas in reality, the same amount of cladding may be used or it may scale differently to the cores. Finally, the effect of compound methods such as interpolation followed by Gaussian smoothing was not considered.

In the chapters that follow, work developing two optical endoscopic imaging techniques based on the PolyScope is described. Both of these techniques aim to detect signal from fluorescence or reflectance, meaning resolution is considered secondary to signal and accuracy of spectral reconstruction, so unless otherwise stated, interpolation based decombing and demosaicing is applied. By using the PolyScope to deliver to and collect light from the oesophagus, it was possible to quickly translate these techniques to first-in-human trials.

References

1. Gaab MR (2013) Instrumentation: endoscopes and equipment. World Neurosurg 79:S14.e11–S14.e21
2. Trindade AJ, Smith MS, Pleskow DK (2016) The new kid on the block for advanced imaging in Barrett's esophagus: a review of volumetric laser endomicroscopy. Therap Adv Gastroenterol 9:408–416
3. NvisionVLE® Imaging System—NinePoint Medical. Available at: http://www.ninepointmedical.com/nvisionvle-imaging-system/. Accessed 1 Aug 2017
4. Rodriguez SA et al (2010) Ultrathin endoscopes. Gastrointest Endosc 71:893–898
5. Saeian K et al (2002) Unsedated transnasal endoscopy accurately detects Barrett's metaplasia and dysplasia. Gastrointest Endosc 56:472–478
6. Sugimoto H et al (2015) Surveillance of short-segment Barrett's esophagus using ultrathin transnasal endoscopy. J Gastroenterol Hepatol (Australia) 30:41–45
7. Tanuma T, Morita Y, Doyama H (2016) Current status of transnasal endoscopy using ultrathin videoscope for upper GI tract in the world. Dig Endosc 28
8. Sami SS et al (2015) A randomized comparative effectiveness trial of novel endoscopic techniques and approaches for Barrett's esophagus screening in the community. Am J Gastroenterol 110:148–158
9. Moriarty JP et al (2017) Costs associated with Barrett's esophagus screening in the community: an economic analysis of a prospective randomized controlled trial of sedated versus hospital unsedated versus mobile community unsedated endoscopy. Gastrointest Endosc
10. Iddan G, Meron G, Glukhovsky A, Swain P (2000) Wireless capsule endoscopy. Nature 405:417
11. Fisher LR, Hasler WL (2012) New vision in video capsule endoscopy: current status and future directions. Nat Rev Gastroenterol Hepatol 9:392–405
12. Wang A et al (2013) Wireless capsule endoscopy. Gastrointest Endosc 78:805–815
13. Fernandez-Urien I, Carretero C, Armendariz R, Muñoz-Navas M (2008) Esophageal capsule endoscopy. World J Gastroenterol 14:5254

14. Ciuti G, Menciassi A, Dario P (2011) Capsule endoscopy: from current achievements to open challenges. IEEE Rev Biomed Eng 4:59–72
15. Gora MJ et al (2013) Imaging the upper gastrointestinal tract in unsedated patients using tethered capsule endomicroscopy. Gastroenterology 145:723–725
16. Liao Z, Gao R, Xu C, Xu D-F, Li Z-S (2009) Sleeve string capsule endoscopy for real-time viewing of the esophagus: a pilot study (with video). Gastrointest Endosc 70:201–209
17. Gora MJ et al (2013) Tethered capsule endomicroscopy enables less invasive imaging of gastrointestinal tract microstructure. Nat Med 19:238–240
18. Gupta N et al (2012) Longer inspection time is associated with increased detection of high-grade dysplasia and esophageal adenocarcinoma in Barrett's esophagus. Gastrointest Endosc 76:531–538
19. Gkolfakis P, Tziatzios G, Dimitriadis GD, Triantafyllou K (2017) New endoscopes and add-on devices to improve colonoscopy performance. World J Gastroenterol 23:3784–3796
20. Gralnek IM et al (2014) Standard forward-viewing colonoscopy versus full-spectrum endoscopy: an international, multicentre, randomised, tandem colonoscopy trial. Lancet Oncol 15:353–360
21. Hassan C et al (2017) Full-spectrum (FUSE) versus standard forward-viewing colonoscopy in an organised colorectal cancer screening programme. Gut 66:1949–1955
22. Clancy NT et al (2012) Multispectral image alignment using a three channel endoscope in vivo during minimally invasive surgery. Biomed Opt Express 3:2567–2578
23. Winter C et al (2006) Automatic adaptive enhancement for images obtained with fiberscopic endoscopes. IEEE Trans Biomed Eng 53:2035–2046
24. Waterhouse DJ, Luthman AS, Bohndiek SE (2017) Spectral band optimization for multispectral fluorescence imaging, vol. 10057, p 1005709
25. Luthman S, Waterhouse D, Bollepalli L, Joseph J, Bohndiek S (2017) A multispectral endoscope based on spectrally resolved detector arrays, p 104110A
26. Regeling B et al (2016) Hyperspectral imaging using flexible endoscopy for laryngeal cancer detection. Sensors 16:1288
27. Elter M, Rupp S, Winter C (2006) Physically motivated reconstruction of fiberscopic images. Proc Int Conf Pattern Recogn 3:599–602
28. Rupp S, Elter M, Winter C (2007) Improving the accuracy of feature extraction for flexible endoscope calibration by spatial super resolution. In: Annual international conference of the IEEE engineering in medicine and biology—proceedings, pp 6565–6571
29. Rupp S et al (2009) Evaluation of spatial interpolation strategies for the removal of comb-structure in fiber-optic images. In: Annual international conference of the IEEE engineering in medicine and biology—proceedings, pp 3677–3680
30. Lee CY, Han JH (2013) Integrated spatio-spectral method for efficiently suppressing honeycomb pattern artifact in imaging fiber bundle microscopy. Opt Commun 306:67–73
31. Han J-H, Lee J, Kang JU (2010) Pixelation effect removal from fiber bundle probe based optical coherence tomography imaging. Opt Express 18:7427
32. Wang P et al (2018) Fiber pattern removal and image reconstruction method for snapshot mosaic hyperspectral endoscopic images. Biomed Opt Express 9:780
33. Pelli DG, Bex P (2013) Measuring contrast sensitivity. Vision Res 90:10–14

Chapter 4
Flexible Endoscopy: Optical Molecular Imaging

4.1 Molecular Imaging for Endoscopic Surveillance of Barrett's Oesophagus

Targeted molecular imaging agents [1] require synthesis at good manufacturing practice (GMP) standards, associated toxicology studies, specific instrumentation, and add time and cost to procedures (Translational Characteristic 1) [2, 3]. Despite these challenges, molecular imaging is an attractive technique as it enables macroscopic visualisation of complex biochemical processes [1]. That is, rather than probing some macroscopic proxy of underlying biochemical change as does HD-WLE or NBI, molecular imaging allows a specific intrinsic molecular change associated with disease to be probed, meaning it has the potential to allow highly specific red flag surveillance of the oesophagus. Efforts in the development of molecular imaging contrast agents for early detection of dysplasia in Barrett's oesophagus are summarised in Table 4.1.

Application of intravenously administered probes is limited as they require at least 24 h between injection and endoscopy, hindering clinical workflow [8, 10]. Encouragingly, labels that can be topically applied during the endoscopic procedure have begun to emerge. A fluorescently labelled peptide that binds specifically to high-grade dysplasia and adenocarcinoma has recently been shown by Sturm et al. to delineate these pathologies in vivo in patients [4], however, the sensitivity to low-grade dysplasia has yet to be established. A follow-up study by Joshi et al, using a wide-field fluorescence imaging endoscope, noted some important limitations, including the challenge in visualising the target fluorescence due to strong tissue autofluorescence, and the resulting low values of target-to-background ratio from the diagnostic fluorescence features [5]. Molecular imaging probes have also been combined with endomicroscopy in several studies [11, 12], but these techniques have a narrow field of view, limiting them to performing optical biopsy rather than red flag imaging.

D. J. Waterhouse, *Novel Optical Endoscopes for Early Cancer Diagnosis and Therapy*, Springer Theses, https://doi.org/10.1007/978-3-030-21481-4_4

Table 4.1 Optical molecular imaging techniques in trials for surveillance of Barrett's oesophagus

Target	Probe	Optical reporter	Device	Subjects	Application	FOV	Prospect
Cyp-A and cell surface protein CD147 (↑) [4]	Peptide: ASYNYDA	FITC	GastroFlex type MiniOminiprobe (Mauna Kea Technologies), Custom Device [5]	In vivo	Topical	Narrow/Wide	++
Surface glycans (↓) [6]	Lectin: Wheat Germ Agglutinin (WGA)	AlexaFluor 488	Gastrointestinal videoscope (GID FQ260Z, Olympus), Custom Device [7]	Whole Organ ex vivo	Topical	Wide	++
Cathepsin B (CTSB) (↑) [8]	Activatable Probe: Prosense 750		Custom Device [9]	Mice	IV	Wide	+
HER2 (↑) [10]	Antibody: anti-HER2	AlexaFluor 488	Cellvizio Confocal CholangioFlex (Mauna Kea Technologies)	Rats	IV	Narrow	–
EGFR, Survivin (↑) [11]	Antibodies: anti-human-EGF-R, anti-human-survivin	Fluorescein	Cellvizio GastroFlex UHD (Mauna Kea Technologies)	Pigs	Topical/IV	Narrow	–
Unknown [12]	Peptide: SNFYMPL	FITC	Axioskop 2 plus microscope (Zeiss)	EMRs ex vivo	Topical	Narrow	–

↑: up-regulated/over-expressed, ↓: down-regulated/under-expressed, γ-Glut: γ-Glutamyltranspeptidase, FITC: fluorescein isothiocyanate, EGFR: epidermal growth factor receptor, HER2: human epidermal growth factor receptor 2, Cyp-A: cyclophilin A, EMRs: endoscopic mucosal resections, IV: intravenous

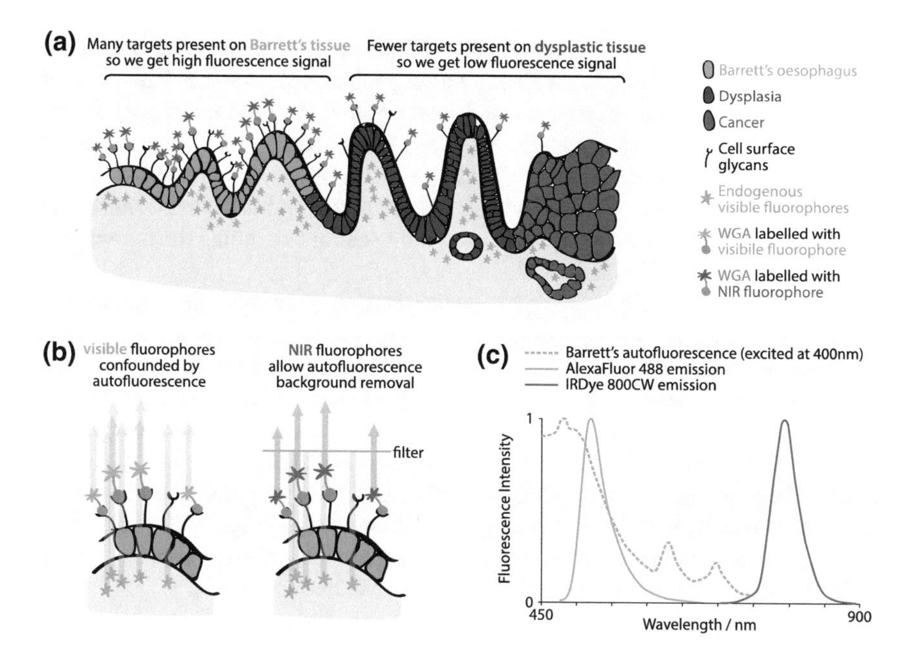

Fig. 4.1 Optical molecular imaging with wheat germ agglutinin. **a** Wheat germ agglutinin (WGA) shows specific changes in binding patterns between high-grade and low-grade dysplastic tissue and non-dysplastic Barrett's oesophagus [6]. **b** Imaging of WGA labelled with visible fluorophores, such as AlexaFluor 488, is confounded by autofluorescence. **c** Spectra show the overlap between autofluorescence and AlexaFluor 488 emission. IRDye 800CW has its emission peak in the near-infrared, far from the autofluorescence of the Barrett's tissue. Autofluorescence data reproduced from [14]. The two small peaks around 630 and 700 nm are due to low dose injection of photofrin. AlexaFluor 488 and IRDye 800CW data from Chroma and LI-COR respectively [16, 17]

4.2 The Potential of Fluorescently Labelled Lectins

Bird-Lieberman et al. demonstrated that cell-surface glycans are altered in the progression from Barrett's oesophagus to adenocarcinoma (Fig. 4.1a). The lectin wheat germ agglutinin (WGA) showed specific changes in binding patterns between high-grade and low-grade dysplastic tissue and non-dysplastic Barrett's oesophagus [6]. WGA has great potential for clinical translation as an OMI contrast agent (Translational Characteristic 1): it can be applied topically using a spray catheter; it requires only 10 min incubation time, minimising the disruption to the normal clinical workflow; it is easily displaced by washing with an excess of glucosamine following imaging; it is inexpensive, stable at high temperature, stable at low pH and resistant to proteolysis; due to its large size (37,300 Da), its binding is less susceptible to alteration once it is conjugated to a fluorescent dye; and lectins are part of a normal human diet and thus have low toxicities if used at sufficiently low concentrations [6].

Next, Bird-Lieberman et al. conjugated WGA to AlexaFluor 488, a commercially available fluorescent dye with excitation peak around 488 nm. This was topically applied to resected oesophagi and imaged in a way that mimics an in vivo clinical study using a commercially available fluorescence endoscope (excitation 395–475 nm, detection 500–630 nm). Finally, the fluorescence signal was correlated to pathology. The results were promising but the signal to background ratios were low in some cases due to a high level of autofluorescence swamping the fluorescence signal from the lectin (Fig. 4.1b, c) [6].

Progression of several promising molecular imaging techniques have encountered this same limitation. Existing endoscopy devices for detecting fluorescence emissions are optimised for detection of tissue autofluorescence in the visible wavelength range [13], meaning that the sensitivity, contrast and dynamic range for detection of exogenously applied contrast agents is limited by the high endogenous background signal present within the images. To avoid the autofluorescence background, imaging can be performed using near-infrared (NIR) fluorophores in the range ~600–900 nm, where the tissue is relatively devoid of endogenous fluorescence (Fig. 4.1c) [14]. Therefore, WGA was conjugated to a commercially available NIR fluorophore, IRDye 800CW-NHS ester dye (LI-COR, USA), to form WGA-IR800 [15].

4.3 Challenges in Realising the Potential of WGA-IR800

Following the identification of a promising OIB, referring back to the OIB Roadmap (Fig. 1.3), it needs to be established if a device exists to measure OIB. While several groups have developed flexible endoscopic devices for NIR fluorescence imaging in the gastrointestinal tract [8, 18–21], there are currently no commercially available NIR fluorescence endoscopes for wide field surveillance of the oesophagus, so a collaboration to develop a NIR fluorescence endoscope in parallel to WGA-IR800 was initiated.

To enable clinical studies, both the device, and WGA-IR800 must be approved for in vivo trials. With regards to WGA-IR800, it was hoped that the combination of WGA's low toxicology profile and the GMP qualification of IR800 would facilitate this stage of translation. With regards to the device, to simplify approval, the NIR fluorescence endoscope was designed around the PolyScope accessory channel endoscope (Sect. 3.2). Since the PolyScope is CE marked, the NIR fluorescence endoscope requires only local approval by clinical engineering (Translational Characteristic 2).

To meet the experimental aims, the specification for the NIR fluorescence endoscope was as follows: it must be compact and robust enough to be deployed in a clinical environment; it must provide image guidance to the endoscopist so they are able to orient themselves in the oesophageal lumen; and it must be capable of acquiring wide field images of low concentrations of WGA-IR800, which is particularly challenging since most detectors are limited by reduced sensitivity in the NIR.

To meet these specifications, a compact bimodal endoscope was developed. The device is capable of detecting both NIR fluorescence for OMI, using a highly sensitive electron multiplying CCD (EMCCD), as well as traditional white light reflectance to enable guidance, using a second monochrome camera. The system was kept compact through the use of mirrors to bend the light path. With this device, NIR fluorescence imaging of WGA-IR800 has the potential to provide a wide field red flag surveillance technique capable of achieving high sensitivity and specificity for detection of dysplasia, improving the standard of care for Barrett's surveillance. The following sections detail the design, development and validation of this device in Domain 2 of the OIB Roadmap (Fig. 1.3) in preparation for first-in-human trials.

4.4 Materials and Methods

4.4.1 Fluorescent Lectin Synthesis

André Neves (Senior Research Associate, CRUK Cambridge Institute) conjugated IRDye800CW NHS Ester (LI-COR, USA) with WGA (L9640, Sigma-Aldrich, USA) to produce WGA-IR800 according to the methods of Sato et al. [22].

4.4.2 Endoscope Design

The system is based around the PolyScope accessory channel endoscope described in Sect. 3.2 and is shown in Fig. 4.2. Illumination is provided by a broadband halogen light source (OSL2B2, Thorlabs, Germany) clipped with a 750 nm short pass excitation filter (FESH0750, Thorlabs, Germany). Light is directed onto the sample through the light guide embedded inside the PolyScope catheter. Light collected by the imaging fibre is relayed through an infinity corrected 20× objective lens (421350-9970-000, Zeiss, Germany) and split into two channels with a long pass dichroic filter (DMLP650L, Thorlabs, Germany). Light with a wavelength in the range 400–633 nm is relayed by an achromatic doublet (f = 100 mm, ACA254-100-A, Thorlabs, Germany) onto a grayscale CMOS sensor (Grasshopper3 GS3-U3-41C6M-C, PointGrey, USA) whilst light with a wavelength > 685 nm is relayed by an achromatic doublet (f = 100 mm, ACA254-100-B, Thorlabs, Germany) onto an electron multiplying CCD (EMCCD; ProEM+ eXcelon 512 × 512, Princeton, USA) through an 800 nm long pass emission filter (FELH0800, Thorlabs, Germany) allowing simultaneous white light reflectance and NIR fluorescence imaging.

The system is kept compact through the use of broadband mirrors (BB1-E02 and BB1-E03, Thorlabs, Germany) to bend the light path. The optics are securely housed inside a light tight enclosure and mounted on an optical breadboard (MB4545/M, Thorlabs, Germany) which is fixed to a stainless steel trolley (FW2901-3, Freeway

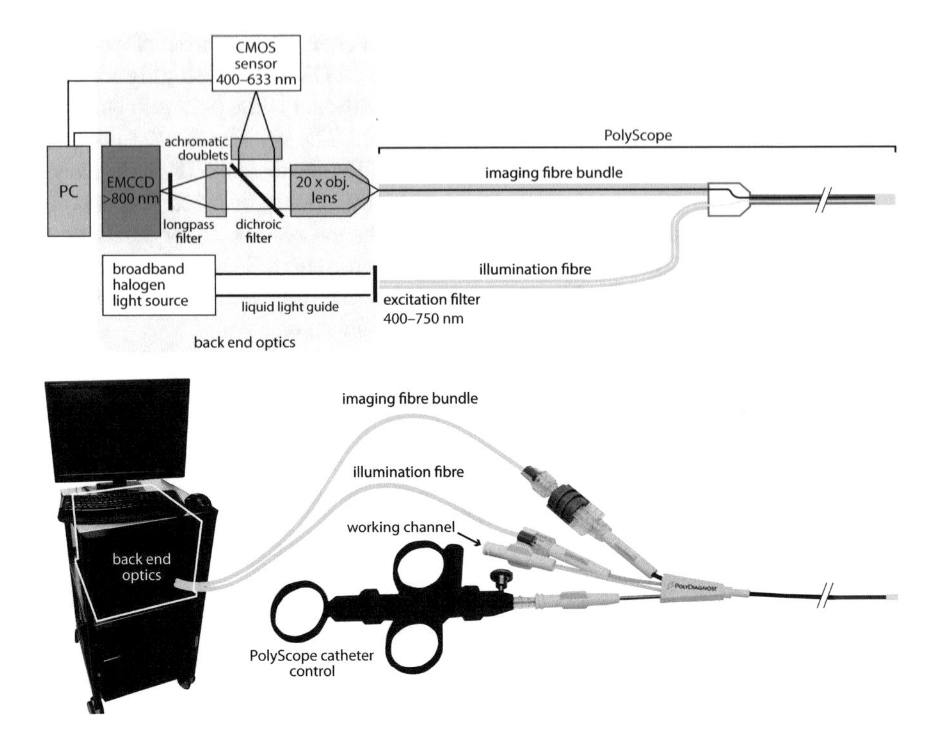

Fig. 4.2 Schematic of the PolyScope based bimodal NIR fluorescence and WL reflectance endoscope. The system is based around the PolyScope accessory channel endoscope (PolyDiagnost, Germany). Illumination is provided by a broadband halogen light source (OSL2B2, Thorlabs, Germany). Light from the PolyScope 10,000 fibre imaging bundle is focused by an infinity corrected 20× objective lens (421350-9970-000, Zeiss, Germany) and split into two channels with a long pass dichroic filter (DMLP650L, Thorlabs, Germany). Light with a wavelength in the range 400–633 nm is relayed by an achromatic doublet lens ($f = 100$ mm, ACA254-100-A, Thorlabs, Germany) onto a grayscale CMOS sensor (Grasshopper3 GS3-U3-41C6 M-C, PointGrey, USA) whilst light with a wavelength > 685 nm is relayed by an achromatic doublet ($f = 100$ mm, ACA254-100-B, Thorlabs, Germany) onto an electron multiplying CCD (EMCCD; ProEM + eXcelon 512 × 512, Princeton, USA) through an 800 nm long pass emission filter (FELH0800, Thorlabs, Germany) allowing simultaneous white light reflectance and NIR fluorescence imaging

Medical, UK) with a footprint of 512 mm × 480 mm. This allows the system to be easily and safely transported around a crowded endoscopy suite (Translational Characteristic 2).

4.4.3 Image Acquisition and Image Corrections

A LabVIEW (National Instruments, USA) Visual Interface (VI) was created to control the cameras and acquire images. The white light images are saved as 8-bit 2048 × 2048 TIFF files and the NIR images were saved as 16-bit 512 × 512 TIFF files.

Data analysis was carried out using Matlab® (MathWorks, USA). The comb structure introduced by the fibre bundle was removed using the interpolation methods outlined in Sect. 3.3.3. The white light and NIR images can be coregistered using the fibre locations, by selecting the four fibres located at the extreme x and y positions. The images are cropped at these locations and a similarity transformation is generated to successfully transform the white light image in x and y such that the extreme fibres are coregistered with the equivalent extreme fibres in the NIR image.

4.5 Technical Characterisation

Prior to translating a technique to the clinic, it is important to ensure it is well characterised, both with respect to its safety, for example by quantifying its illumination power (Sect. 4.5.1.2), and with respect to its intended operation, for example by quantifying its field of view (Sect. 4.5.1.1), resolution (Sect. 4.5.1.3) and sensitivity (Sect. 4.5.1.4). Whilst most of these are properties of the device, note that sensitivity is a combined characteristic of the device and molecular imaging probe.

4.5.1 Methods

4.5.1.1 Field of View

To enable wide field surveillance of the whole oesophagus it is important that the endoscope has sufficient imaging performance across a wide field of view (FOV). In order to measure the FOV we captured 3 images of 1 mm graph paper at 4 different working distances (WD), 5.7, 8.7, 11.7 and 16.7 mm (error $\pm$ 0.3 mm). The resulting images are expected to show barrel distortion defined by:

$$r_u = Ar_d\left(1 + kr_d^2\right) \tag{4.1}$$

where r_u is the radial distance from the center of the ground truth image to a given vertex i in mm, r_d is the radial distance from the centre of the distorted image to the same vertex i in pixels, k is a constant that describes the magnitude of the distortion and A is a constant used to convert between units of pixels and mm. For each of the 12 images acquired (4 working distances $\times$ 3 replicates) the position of the centre of the fibre bundle was identified and the radial distance to each vertex in the graph paper, r_d (in pixels), was plotted against the true distance, r_u (in mm), which is known from measuring the physical distance between the markings on the graph paper.

4.5.1.2 Power

In order for a device to be approved for first-in-human trials, its maximum power must be below the threshold above which thermal damage, photosensitisation and photoallergic reactions may occur [23]. However, the published guidelines for optical radiation exposure limits only concern the retina and skin, and are limited in their application to workers rather than patients, so are not relevant when determining the maximum permissible exposure (MPE) for endoscopic imaging [23, 24] (Translational Barrier 1). Therefore, MPE was established by measuring the maximum broadband power of a clinical endoscope (Olympus Evis Lucera CVL-260SL Xenon light source with an Olympus Gastroscope GIF-FQ260Z, Japan) in normal operating conditions, using a thermal power meter (spectral range 0.19–2.1 μm; A-02-D12-BBF-USB, LaserPoint, Italy). Since this standard of care device is approved for clinical use in humans, it can be assumed that this MPE represents safe illumination levels for endoscopic use. The power measurement was repeated for the bimodal endoscope, and the results compared to ensure the determined power of this device was less than the power of the standard of care endoscope.

4.5.1.3 Resolution

The resolution of the system is fundamentally limited by the fibre bundle structure. To determine the limiting resolution, images of a 1951 USAF resolution test target (#53-714, Edmund Optics, USA) were captured at 4 working distances using external illumination from a broadband halogen light source (OSL2B2, Thorlabs, Germany) to reduce specular reflections. Background subtraction was performed using an averaged dark image of 10 frames acquired at the end of the experiment. White light images were analysed since the test target is printed on white reflective photo paper intended for visible imaging. Since both imaging channels are able to resolve the individual fibres of the fibre bundle, the limiting resolution is determined by the fibre bundle properties rather than the properties of the cameras, and so the resolution measured using the white light images is applicable to both imaging channels.

4.5.1.4 Sensitivity

For in vivo application, it is important that WGA-IR800 can be detected at low concentrations. The reasons for this are threefold. Firstly, since binding of WGA-IR800 decreases with the progression of the disease, detecting it at low concentrations is important to avoid false positives. Secondly, minimising the amount of any exogenous agent sprayed inside the oesophagus is desirable for clinical translation. Thirdly, it is desirable to minimise non-specific binding effects that may become most problematic at high concentrations where they have the potential to reduce contrast. Therefore, determining the sensitivity of the system is a crucial aspect of the characterisation process.

In order to characterise sensitivity, a two-fold dilution series of WGA-IR800 in phosphate buffered saline (PBS) was prepared and 30 μL of each solution was pipetted into a well plate (μ-Slide 18 Well-Flat, ibidi, Germany), which had been spray-painted matte black to avoid specular reflections. Images of the dye were captured at 5 different working distances (5, 10, 12, 16 and 20 mm) representative of the range of distances that would be used in vivo. The NIR images were captured using an EM gain of 20 and an exposure time of 200 ms. The images were corrected and coregistered as described in Sect. 4.4.3. 50 × 50 pixel regions of interest (ROIs) were drawn on the white light images inside (signal) and outside (background) of the well. These ROIs were then applied to the coregistered NIR images for analysis in Matlab. The signal-to-noise ratio (SNR) was calculated using:

$$SNR = \left(\frac{S - B}{\sigma}\right) \tag{4.2}$$

where S is the centre of a Gaussian fitted to the pixel intensity distribution in the signal ROI, B is the mean of a Gaussian fitted to the pixel intensity distribution in the background ROI and σ is the standard deviation of a Gaussian fitted to the pixel intensity distribution in the background ROI.

The relationship between the SNR and electron multiplying (EM) gain of the NIR sensor was also characterised by capturing images of 30 μL of 3100 nM WGA-IR800 while varying the gain at working distances of 5, 9.5 and 12 mm. For an EMCCD, the expected relationship between SNR and gain is:

$$SNR \propto \left(constant + \frac{1}{G^2}\right)^{-\frac{1}{2}} \tag{4.3}$$

where G is the EM gain of the sensor [25].

4.5.2 Results

4.5.2.1 Field of View

First, the field of view (FOV) of the system was assessed. An example image displaying barrel distortion can be seen in Fig. 4.3. The distortion constant k and the constant A were determined by fitting Eq. (4.1) to the data ($R^2 = 0.9949$–0.9974). The values of k and A were then used to determine the FOV radius ($= r_u$) based on the radius of the images in pixels ($= r_d$). Combining these data for the 4 working distances, we determined the angle of the FOV to be 63° ± 1°, which compares favourably to the manufacturer specified angle of 70°.

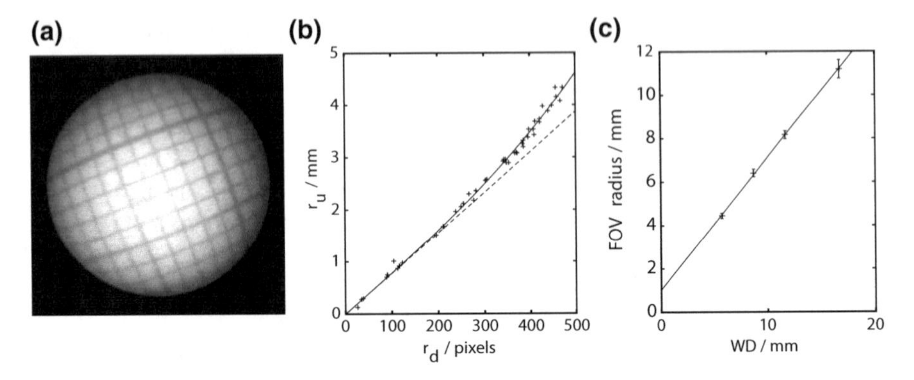

Fig. 4.3 Characterisation of the field of view (FOV) of the bimodal endoscope. **a** An endoscopic image of 1 mm square graph paper shows barrel distortion. From this image, r_u and r_d were measured for several vertices on the paper. **b** An example of the fit to Eq. (4.1) for images taken at a working distance (WD) of 5.7 mm ($R^2 = 0.9954$). The straight dashed line represents the case of no barrel distortion. The fit was used to extract the constant *A* and the distortion parameter *k*. The values of *A* and *k* were used with Eq. (4.1) to determine the FOV radius ($=r_u$) based on the radius of the images in pixels ($=r_d$). **c** Determined FOV radius for four WDs ($R^2 = 0.9996$). Error bars represent the standard error of the FOV radius derived from the standard errors of the fit parameters *A* and *k*. From the fitted line, the angular FOV was calculated to be $63 \pm 1°$

4.5.2.2 Power

To establish a safe maximum power for the endoscope, its maximum broadband power was compared to that of a clinical endoscope. The maximum power from the clinical endoscope in white light reflectance mode was measured to be 19 ± 1 mW at a working distance of 1.0 ± 0.1 cm whilst the maximum power from the bimodal endoscope was measured to be 3.9 ± 0.2 mW at a working distance of 1.0 ± 0.1 cm. Hence it was assumed that the bimodal endoscope was well within the safety limits adhered to by current commercial endoscopes.

4.5.2.3 Resolution

Resolution was determined by taking images of a USAF test target. At least 4 replicate data points for each line pair element were captured (centred within the FOV) each at working distances of 5, 10, 15 and 20 mm. The Michelson contrast (Eq. 3.17) was calculated for each element (after comb artefact correction) and the results plotted against the reciprocal of the line width of a single line element (Fig. 4.4).

A contrast of 1% that has been reported to be the minimum contrast required for the detection of a pattern across a wide range of targets and conditions [26], but a contrast threshold of 5% was chosen to avoid effects arising from noise at very low contrast. By finding the intersect of this threshold with exponential fits, the resolution

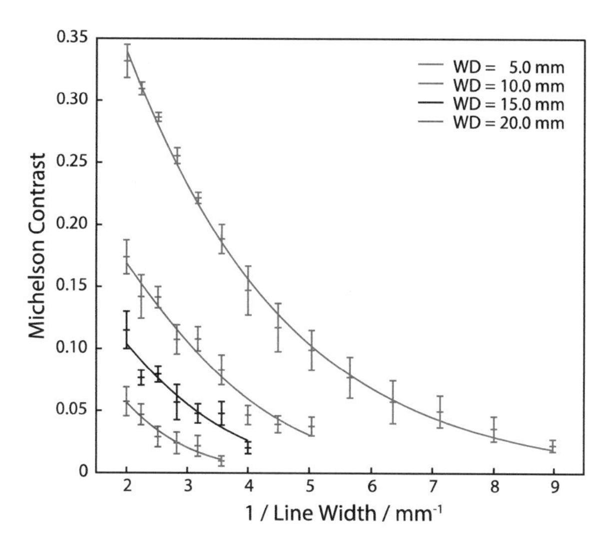

Fig. 4.4 Characterisation of the resolution of the bimodal endoscope. Resolution was characterised by imaging a USAF chart at four different working distances. The resolution was determined as the point where an exponential fit drops below 5% Michelson contrast. R^2 = 0.9965, 0.9665, 0.9068 and 0.9640 for working distances of 5, 10, 15 and 20 mm respectively

of the bimodal endoscope was determined to be 148 ± 7, 230 ± 20, 320 ± 40 and 470 ± 40 μm at working distances of 5, 10, 15 and 20 mm respectively.

4.5.2.4 Sensitivity

Finally, the system sensitivity was established by investigating the dependence of signal-to-noise ratio (SNR) on concentration; a SNR of 3 was used to determine the minimum detectable concentration at each working distance. Example images from the white light and NIR channels are shown in Fig. 4.5a while Fig. 4.5b shows the relationship between SNR and concentration at working distances of 5, 10, and 12 mm. At working distances beyond 16 mm the dye was not detectable at any concentration ≤3100 nM. A linear fit is made to these data for SNR > 1 since below this only noise is expected. For a working distance of 5 mm, the observed non-linear region at high concentration was likely due to saturation, hence was excluded from the linear fit. The minimum detectable concentrations assessed from Fig. 4.5b are 110 ± 60 nM and 430 ± 170 nM for working distances of 5 and 10 mm respectively. Without comb correction, the minimum detectable concentrations were determined as 170 ± 40 nM and 810 ± 520 nM for working distances of 5 and 10 mm respectively, confirming that removal of the comb artefact grants an improvement in SNR, since it removes low signal pixels representing the cladding between fibrelets (Sect. 3.4).

The influence of working distance and EMCCD gain on these data was also assessed. The relationship between sensitivity and working distance arises due to the dependence on the irradiance of the excitation light, which decreases with increased WD. This was investigated by capturing images of 30 μL of 3100 nM WGA-IR800 at 8 working distances and plotting SNR against working distance (Fig. 4.5c). The fit suggests that the SNR falls off as working distance to the power of −1.91 ± 0.08, consistent with the inverse square law expected for an illumination cone. Figure 4.5d shows the fit of the EM gain and SNR data to Eq. (4.3). Increasing EM gain reduces the lifetime of the EMCCD. According to our data, using a gain of ~20 provides maximises SNR whilst minimising the EM gain required to achieve this SNR, preserving the lifetime of our EMCCD.

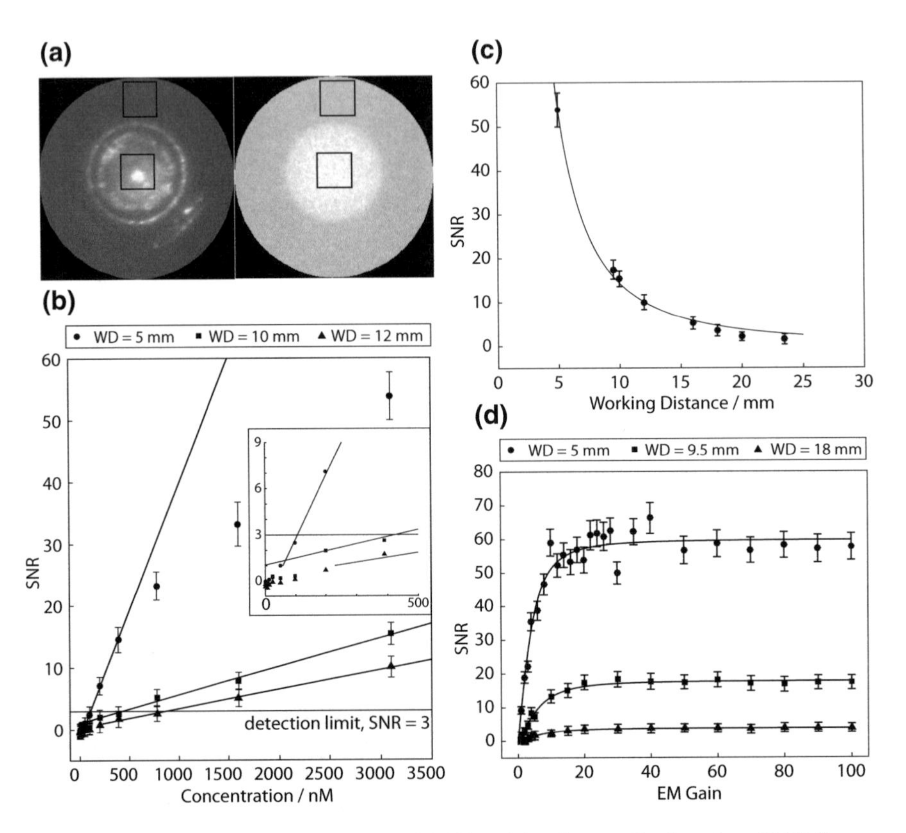

Fig. 4.5 Characterisation of the sensitivity of the bimodal endoscope for detection of NIR fluorescence from WGA-IR800. **a** Coregistered images of white light (left) and NIR fluorescence (right) with example ROIs illustrated for signal and background. **b** The signal to noise ratio (SNR) increases with the concentration of WGA-IR800. Linear fits were applied giving $R^2 = 0.9987$, 0.9936 and 0.9974 for working distances of 5, 10 and 12 mm respectively. The horizontal purple line shows the detection limit defined as SNR = 3. Inset shows a magnification of the region at low concentration. Error bars have been removed for clarity. **c** SNR of 3100 nM WGA-IR800 follows an inverse power law with the working distance: $R^2 = 0.9952$. **d** SNR eventually reaches a plateau with increasing EM gain applied to the EMCCD for images taken of 3100 nM WGA-IR800. The lines show the fit according to Eq. 4.3 with $R^2 = 0.9358$, 0.9819 and 0.9727 for working distances of 5, 9.5 and 18 mm respectively. (Exposure time = 200 ms and EM gain = 20 unless otherwise stated. Error bars represent the standard error of the SNR derived from the standard deviation of pixel values in the signal ROI)

4.6 Validation Using Biological Samples

To validate WGA-IR800 imaging with the bimodal NIR fluorescence and WL reflectance endoscope, a range of biological samples were tested (Domain 2, OIB Roadmap, Fig. 1.3). These experiments are summarised in Table 4.2.

As a preliminary validation of the ability to distinguish between different tissue types using the proposed technique, the novel bimodal endoscope was used to acquire images of healthy ex vivo animal tissue stained with WGA-IR800 (Sect. 4.6.1.1). Mouse stomachs were chosen, as they are readily available, and contain two distinct regions: the upper non-glandular forestomach, which has squamous tissue at the exposed surface, and the lower glandular stomach, which has simple columnar epithelium (gastric type) tissue at the exposed surface. These provide a model of the corresponding tissue types found in the human oesophagus in healthy (squamous) and Barrett's oesophagus (columnar-lined oesophagus) respectively.

In parallel to this work, WGA-IR800 was applied to ex vivo human tissue to ensure that the original results of Bird-Lieberman et al. [6], which validated that WGA-AlexaFluor488 could distinguish between dysplastic tissue and Barrett's oesophagus, were reproducible with WGA-IR800 (Sect. 4.6.1.2). Since tissue begins to degrade once it is removed from the body, ex vivo tissue does not usually provide a good model of the contrast expected in vivo (Translational Barrier 4). However, the cell surface sialic acid residues targeted by WGA-IR800 should remain intact for some time after resection [6], so experiments using ex vivo tissue proceeded here.

Human tissue was taken via endoscopic mucosal resection, a procedure which removes a ~2 cm diameter section of superficial tissue, called an endoscopic mucosal resection (EMR), from the lining of the oesophagus. A gold standard NIR fluorescence imaging system (Fluobeam-800, Fluoptics, France), which is compact enough to be placed directly in the endoscopy suite, was used to image freshly collected EMRs, and the images were compared to gold standard disease classification from

Table 4.2 Experiments used to validate WGA-IR800 imaging with the bimodal NIR fluorescence and WL reflectance endoscope

Section	Objective	Sample	Imaging comparator 1	Imaging comparator 2	Pathology comparator
4.6.1.1	Preliminary validation of the ability of NIR fluorescence endoscope + WGA-IR800 to distinguish between different tissue types	Ex vivo mouse stomach	NIR fluorescence endoscope + WGA-IR800	None	Squamous and gastric tissue are distinguishable by eye
4.6.1.2	Validation of the ability of WGA-IR800 to distinguish between dysplasia and Barrett's oesophagus	Ex vivo human endoscopic mucosal resection specimens (EMRs)	Wide field fluorescence imaging device + WGA-IR800	None	Histopathology of punch biopsies
4.6.1.3	Validation of the accuracy of the NIR fluorescence endoscope for detection of WGA-IR800		NIR fluorescence endoscope + WGA-IR800	Wide field fluorescence imaging device + WGA-IR800	None
4.6.1.3	Validation of the ability of NIR fluorescence endoscope + WGA-IR800 to distinguish between different disease types		NIR fluorescence endoscope + WGA-IR800	None	Histopathology of entire resection specimen

Gold coloured cells represent gold standards

histopathology. The advantage of this on-site method is twofold: first, since the samples are fresh, the degradation of tissue is minimised; second, the ease of the procedure and proximity to available specimens allows a large sample size to be achieved.

Next, the novel endoscope was used to capture images of WGA-IR800 stained EMRs (Sect. 4.6.1.3). To evaluate the accuracy of the novel device for detection of WGA-IR800, the images were compared to those captured using the gold standard imaging device. Finally, images of WGA-IR800 stained EMRs captured with the novel endoscope were compared to gold standard disease classification from histopathology to validate the ability of the technique to distinguish between disease types (Sect. 4.6.1.3).

4.6.1 Methods

4.6.1.1 Imaging of Mouse Stomachs

To demonstrate the feasibility of imaging WGA-IR800 on a tissue background, and the ability to distinguish between different tissue types, excised mouse stomachs were stained with the fluorescently labelled lectin and images were acquired using the bimodal endoscope. Surplus mice from breeding were obtained post mortem ($n = 6$), so were not classed as regulated procedures under the UK Animals (Scientific Procedures) Act, 1986.

Mouse stomachs were prepared by opening and pinning the excised stomachs on parafilm covered cork. The stomachs were washed with PBS to remove any contents before the first round of imaging. This washing procedure consisted of tilting the stomachs and pouring the PBS such that it ran from left to right over the stomachs to avoid cross contamination between the tissue types (Fig. 4.7b). Data acquisition was then performed using EM gain = 20 at WD = 7 ± 1 mm. After the baseline imaging, 100 μL of 29 μg/mL (780 nM) WGA-IR800 was evenly pipetted onto the face of the stomachs and left to incubate for 10 min. The stomachs were then imaged four more times: immediately after incubation; after a wash with 5 mL of PBS; after a further wash of 15 mL of PBS; and after a final wash of 45 mL PBS. Each round of imaging captured three images: squamous tissue; gastric tissue; and parafilm covered cork to serve as a background control. For each image, an ROI was drawn in the centre of the image and the mean signal calculated by fitting a Gaussian to the pixel intensity distribution.

In order to confirm the differential binding of WGA to squamous and gastric tissue types, we repeated the experiment using an additional wash with 33 mL of 1 mM *N*-acetyl-D-glucosamine (Sigma-Aldrich) in PBS, which has been shown previously to compete with WGA for sialic acid binding [6]. The washing was performed in the same manner as described above. Excess glucosamine that did not run off was left to incubate for 5 min before a final wash of 50 mL of PBS.

4.6.1.2 Imaging of Ex Vivo Human Resection Specimens with a Gold Standard Imaging Device

Together with our collaborators Massimiliano di Pietro and André Neves, an ex vivo study of 29 EMRs was performed to ensure that the original results of Bird-Lieberman et al. [6], which validated WGA-AlexaFluor488 could distinguish between dysplastic tissue and Barrett's oesophagus, were reproducible with WGA-IR800 [15]. Consecutive patients undergoing endoscopic mucosal resection for Barrett's related early neoplasia were recruited. Freshly collected EMR specimens were washed with 5 mL of phosphate-buffered saline (PBS), sprayed with 2 mL of WGA-IR800 (10 μg/mL) in PBS, incubated in the dark for 10 min at room temperature, and then washed again with the same buffer, prior to imaging using a wide field intraoperative fluorescence imaging device (Fluobeam-800, Fluoptics, France). A maximum of two punch biopsies (2 mm diameter discs of tissue) were collected ex vivo from each EMR specimen under NIR fluorescence guidance; one from a high intensity region and one from a low intensity region. These were fixed in formalin and classified by a pathologist according to the Vienna classification [27].

4.6.1.3 Imaging of Ex Vivo Human Resection Specimens with the Bimodal Endoscope

To ensure the accuracy of the bimodal endoscope for detection of WGA-IR800 on human tissue, a pilot study assessing 5 EMRs from a single patient undergoing endoscopic therapy for Barrett's related intramucosal adenocarcinoma was performed. This study was approved by the Cambridgeshire 2 Research Ethics Committee (09/H0308/118).

Resections were first washed with PBS to remove superficial debris and excess mucus then stained with 10 μg/mL (268 nM) WGA-IR800 using a small spray bottle to mimic the spray catheter application that would be used in endoscopic practice. Tissues were left to incubate for 10 min at room temperature then washed with 15 mL of PBS to remove unbound fluorescent probe. The EMRs were first imaged with a wide field intraoperative fluorescence imaging device (Fluobeam-800, Fluoptics, France), representing the gold standard for imaging of NIR fluorescence, and then with the bimodal endoscope.

To fit the entire EMRs (~2.5 cm diameter) into a single field of view, images were captured at a working distance of ~2 cm using the bimodal endoscope, resulting in a decrease in illumination power density at the sample surface and resulting SNR. To counter this, the exposure time was increased to 2 s, to obtain adequate signal in the images.

In order to compare the endoscope images with the gold standard images, the images were coregistered using the following procedure. First, ROIs were manually drawn around the EMR in the gold standard image and the white light endoscope image (which is already coregistered with the NIR endoscope image as described in Sect. 4.4.3). Using these ROIs, binary masks were generated, which were then coregistered using the 'imregtform' function in Matlab, which coregisters images based on mutual information. This generated a similarity transformation that was applied to the raw gold standard image to coregister it with the endoscope images. Finally, using both the endoscope mask and the coregistered gold standard mask, the coregistered images were masked so that only image points inside both of the initial ROIs would be compared in the final coregistered images. This process is summarised in Fig. 4.6.

After NIR imaging, EMR specimens were fixed in formalin and embedded in paraffin, according to standard histopathological procedures at Cambridge University Hospitals Human Research Tissue Bank. The EMR paraffin block was cut at intervals of 2 mm and sections were mounted onto glass slides. Slides were H&E stained and scanned using an Axio Scan.Z1 and imported into Zen 2 lite software (both Carl Zeiss Microscopy, Germany). The EMR sections were scored by the study pathologist every 1 mm for pathological grade according to the Vienna classification [27], with the help of the ZEN software graphical grid. This allowed construction of a pathology grid, which was superimposed manually onto gold standard fluorescence images. To facilitate histopathological correlation, the normal oesophageal squamous epithelium and oesophageal gastric/intestinal metaplasia (non-dysplastic Barrett's) were grouped together as "non-dysplastic" and any grade of neoplasia, including indefinite for dysplasia, low- and high-grade dysplasia and intramucosal cancer, were grouped together as "dysplastic". After processing the 5 EMR samples and excluding artefacts (including pins, edge effects due to pooled dye under the edges of the tissue and ulcers), it was only possible to perform an accurate co-registration between endoscopic images and histology for one EMR sample.

4.6.2 Results

4.6.2.1 Imaging of Mouse Stomachs

To demonstrate the feasibility of imaging WGA-IR800 on a tissue background, and the ability to distinguish between different tissue types, excised mouse stomachs were stained with the dye and data was acquired using the bimodal endoscope. The upper non-glandular forestomach, which has squamous tissue at the exposed surface, and the lower glandular stomach, which has simple columnar epithelium (gastric type) tissue at the exposed surface, provide us with a model of the corresponding tissue types found in the human oesophagus in healthy (squamous) and Barrett's oesophagus (columnar-lined oesophagus) respectively. The different regions of tissue within the stomach are indicated in Fig. 4.7a, b.

Gastric tissue was clearly distinguishable (2 way ANOVA, $p = 0.0005$) from the squamous tissue using WGA-IR800 and the bimodal endoscope (Fig. 4.7c), even following extensive washing with PBS, suggesting that the WGA-IR800 binds strongly to the gastric mouse tissue. The results shown in Fig. 4.7d show a significant reduction of the fluorescence to background level (2-way ANOVA, $p = 0.0472$)

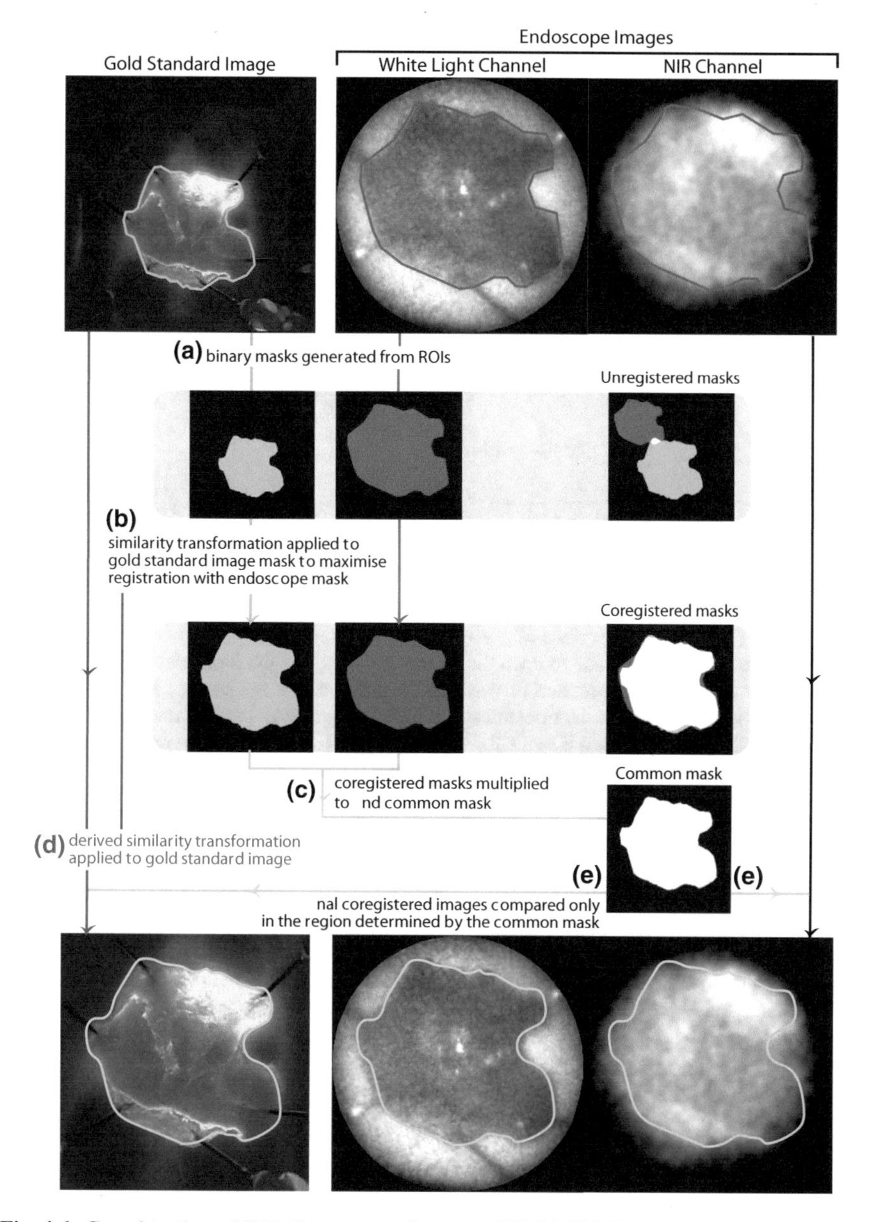

Fig. 4.6 Coregistration of NIR fluorescence images of WGA-IR800 binding to human EMRs. **a** ROIs drawn around the EMRs in the Fluobeam image and white light endoscope image are turned into binary masks. Initially these are unregistered. **b** The Fluobeam mask is similarity transformed in order to maximise overlap with the endoscope mask. **c** The resulting Fluobeam mask and the endoscope mask are multiplied together to find the common mask. **d** The similarity transform found by coregistering the binary masks is now applied to the Fluobeam image to coregister it with the endoscope image. **e** The coregistered images can be compared in the region defined by the common mask, which represents image points that were found inside both of the original unregistered ROIs

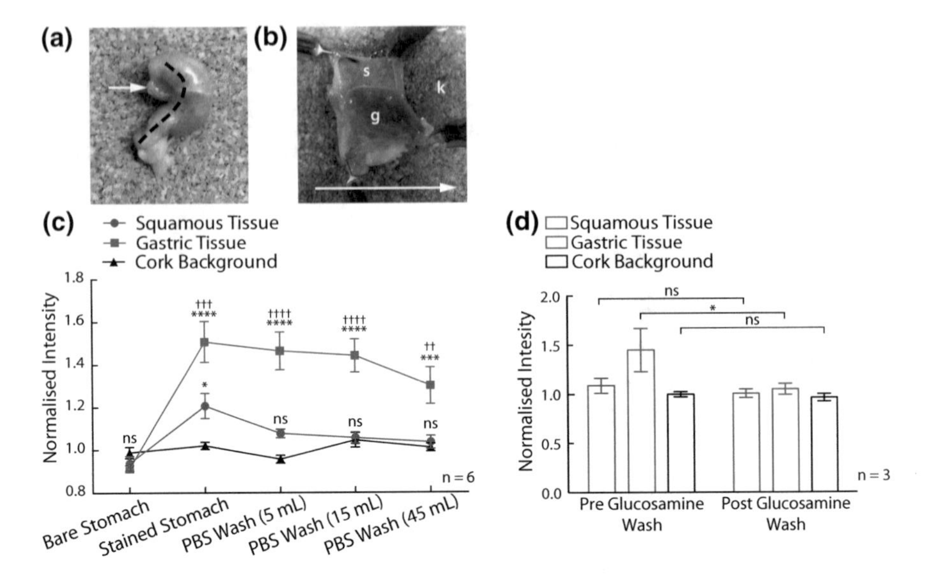

Fig. 4.7 WGA-IR800 binding to excised mouse stomach. **a, b** Photographs of the tissue specimens. The arrow in (**a**) shows the location of the oesophagus. The dotted line shows the approximate location of the cut that was made to open the stomach and expose the inner wall as shown in (**b**). The arrow in (**b**) shows the direction of washing. In both images, we can see the limiting ridge separating the upper non-glandular forestomach, which has squamous tissue at the exposed surface, and the lower glandular stomach is evident. For each stomach three images were taken: one at the centre of the squamous tissue (s); one at the centre of the gastric region (g); and one of cork as a background (k). **c** The stomachs were stained with WGA-IR800 and washed with PBS. At each time point, images of each tissue type and cork were taken and the mean intensity in a central ROI of each was calculated. The intensity was normalised to the average background cork level for each mouse. The mean of n = 6 mice is plotted with error bars representing the standard error in the mean. Statistical testing was carried out using 2-way ANOVA. (ns = no significant difference vs. cork; $*p < 0.05$ versus cork; $***p < 0.001$ versus cork; $****p < 0.0001$ versus cork; $\dagger\dagger\ p < 0.01$ versus squamous; $\dagger\dagger\dagger p < 0.001$ versus squamous; $\dagger\dagger\dagger\dagger p < 0.001$ versus squamous). **d** Following a further wash with glucosamine the bound WGA is removed and the gastric tissue fluorescence intensity returns to the level of the background. Statistical testing was carried out using 2-way ANOVA. (ns = no significant difference; $*\ p < 0.05$)

following the glucosamine wash, confirming the results are due to the differential binding of WGA to different tissue types.

4.6.2.2 Imaging of Ex Vivo Human Resection Specimens with a Gold Standard Imaging Device

Together with our collaborators Massimiliano di Pietro and André Neves, an intra-operative fluorescence imaging device was used to capture images of WGA-IR800 stained EMRs to validate of the ability of WGA-IR800 to distinguish between dysplasia and Barrett's oesophagus. Mean fluorescence intensity was calculated for areas targeted by punch biopsies. To facilitate histopathological correlation, low- and high-grade dysplasia were grouped together as "dysplastic". Normal squamous and intramucosal cancer biopsies were not included in the analysis.

Dysplastic regions had significantly lower mean fluorescence intensity than biopsies collected from regions with non-dysplastic Barrett's epithelium ($p < 0.001$, two-tailed Wilcoxon matched-pairs signed rank test) (Fig. 4.8a). A receiver operating characteristic analysis, with a threshold of mean fluorescence intensity = 0.1115, indicated an area under the curve (AUC) of 0.84 ± 0.07 with a sensitivity of 80% and a specificity of 82.9% for dysplasia (Fig. 4.8b) [15].

4.6.2.3 Imaging of Ex Vivo Human Resection Specimens with the Bimodal Endoscope

To ensure the accuracy of the bimodal endoscope for detection of WGA-IR800, imaging of EMRs collected from a patient with Barrett's oesophagus was performed using the bimodal endoscope and a gold standard imaging system. The coregistered gold standard and endoscope EMR images for all collected specimens are shown in Fig. 4.9a, b respectively. The high signal observed at the edge of the specimens is due to pooling of the dye between the edge of the tissue specimen and the underlying parafilm, which remains even after washing. The intensity recorded in each coregis-

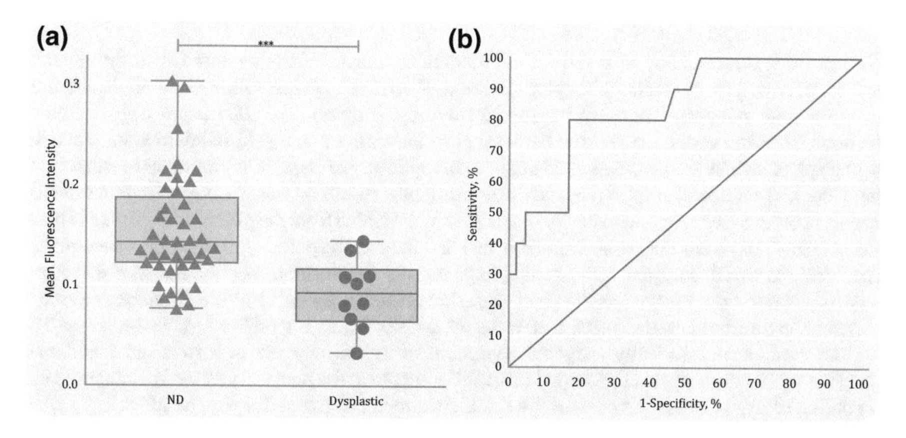

Fig. 4.8 NIR fluorescence intensity correlates with dysplasia in punch biopsies. Consecutive patients (n = 21) undergoing endoscopic mucosal resection (EMR) for Barrett's related early neoplasia were recruited. Freshly collected EMR specimens were sprayed at the bedside with WGA-IR800 and then imaged using an intraoperative fluorescence imaging device (Fluobeam-800, Fluoptics, France). Punch biopsies were collected from each EMR under NIR light guidance. **a** Mean fluorescence intensity for non-dysplastic (ND) punch biopsies is 0.154 ± 0.054 versus 0.092 ± 0.035 for dysplastic biopsies. Normal squamous and intramucosal cancer biopsies were not included in the analysis. n = 51 biopsies, with 10 biopsies (19.6%) containing dysplasia. ***$p < 0.001$. **b** Receiver operating characteristic (ROC) for data in (**c**). Area under the ROC = 0.84 ± 0.07; Sensitivity = 80%; Specificity = 82.9%. Reproduced from [15]

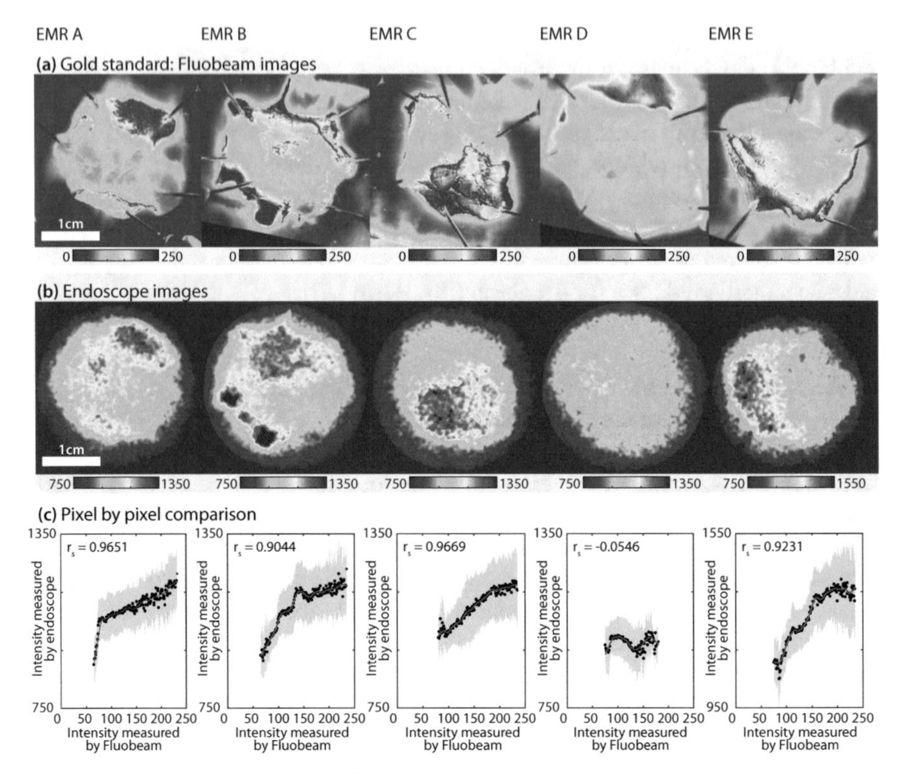

Fig. 4.9 NIR fluorescence of WGA-IR800 binding to human EMRs captured with the bimodal endoscope and the gold standard imaging device. **a** Wide field high resolution NIR images acquired using the gold standard Fluobeam intraoperative imaging system. **b** NIR images acquired using the bimodal endoscope. **c** Correlation between pixel intensities in the gold standard images and the average pixel intensities of coregistered pixels in the endoscope image. The grey areas correspond to the standard deviation of pixels in the endoscope images. A threshold was placed on the gold standard images in order to remove low intensity pixels (due to the signal from the pins holding tissue in place rather than signal from the tissue surface itself) and their corresponding endoscope image pixels. The threshold was determined by removing low intensity Fluobeam pixels until these pixels began to correspond to tissue as well as the pins. The same pixels were then removed from the endoscope images. The thresholds were determined to be 65, 65, 80, 75, 75 for EMRs A–E respectively. The red line shows a robust locally weighted regression smoothing. Spearman correlation coefficients are given within the graphs. High signal intensity observed at the periphery of the specimens is due to pooling of dye between the edge of the tissue specimen and the underlying parafilm

tered pixel in the gold standard and endoscope images was compared to determine if there was correlation between the images and hence whether the data acquired with the endoscope faithfully recapitulates that acquired with the gold standard system.

Intensity scatter plots of these data clearly show a direct relationship between fluorescence intensity in the gold standard and endoscope images for 4 of the 5 EMRs (Fig. 4.9c). Spearman's rank correlation coefficient reveals a moderate but significant correlation between the gold standard and endoscope images in these 4 of 5 EMRs

($r_s = 0.90–0.97$), suggesting WGA-IR800 NIR fluorescence is accurately measured by our endoscope. The lack of significant correlation in EMR D can be explained by the relatively uniform fluorescence observed in both gold standard and endoscope images (fluorescence signal in central 80% of EMR within gold standard images: 110 ± 50, 130 ± 40, 160 ± 40, 100 ± 9, 140 ± 30 for EMRs A–E respectively).

Using EMR B, which had two large regions of non-dysplastic and dysplastic tissue, the NIR endoscopy data was manually coregistered to the pathology grid (Fig. 4.10a). The expected negative binding relationship between dysplastic and non-dysplastic tissue was confirmed (Fig. 4.10b). These results provide a promising preliminary indication that WGA-IR800 fluorescence imaging with the bimodal endoscope is capable of distinguishing between disease pathologies in oesophageal tissue.

4.7 Discussion and Conclusions

To address the clinical challenge of Barrett's surveillance, a bimodal endoscope capable of acquisition and coregistration of white light reflectance images for endoscopic guidance and of NIR fluorescence images for optical molecular imaging of WGA-IR800, a fluorescently labelled lectin that shows differential binding to Barrett's and dysplastic tissue [6], was designed and built. This endoscope is compact, robust and thus compatible with the clinical environment; it can be used alongside standard of care procedures via the working channel of standard endoscopes; it has safe illumination levels and is based on a CE marked device, facilitating local safety approval for use in humans; and it is able to detect WGA-IR800 at concentrations as low as 110 ± 60 nM at the shortest working distances, with a field of view of $63° \pm 1°$ and an image resolution of 141 ± 7 μm.

For preliminary biological validation, the ability of the endoscope to discriminate gastric- from squamous-type tissue in healthy ex vivo mouse stomachs stained with WGA-IR800 was demonstrated ($p < 0.001$). In parallel to this work, the ability to discriminate dysplastic from non-dysplastic tissue was confirmed using WGA-IR800 stained human resection specimens and a gold standard NIR fluorescence imaging system ($p < 0.001$). Finally, the novel bimodal endoscope was compared to a gold standard imaging system for imaging WGA-IR800 stained human resection specimens. The results showed an encouraging correlation between fluorescence signal intensities recorded with the two systems on a per-pixel basis ($r_s = 0.90–0.97$), with this relating directly to histopathological outcome.

For future ex vivo work, it is necessary to overcome remaining coregistration challenges to enable better comparison of fluorescence data to the gold standard histopathological analysis. Inaccuracies in the coregistration of the pathology grid and fluorescence image may arise for several reasons including: unavoidable deformations and artefacts in the processing of EMRs; the assignment of a single majority pathological grade to a large 1×2 mm area, which may contain mixed pathologies and cancer field effects; and the manual coregistration of the pathology grid with the fluorescence image (Translational Barrier 3).

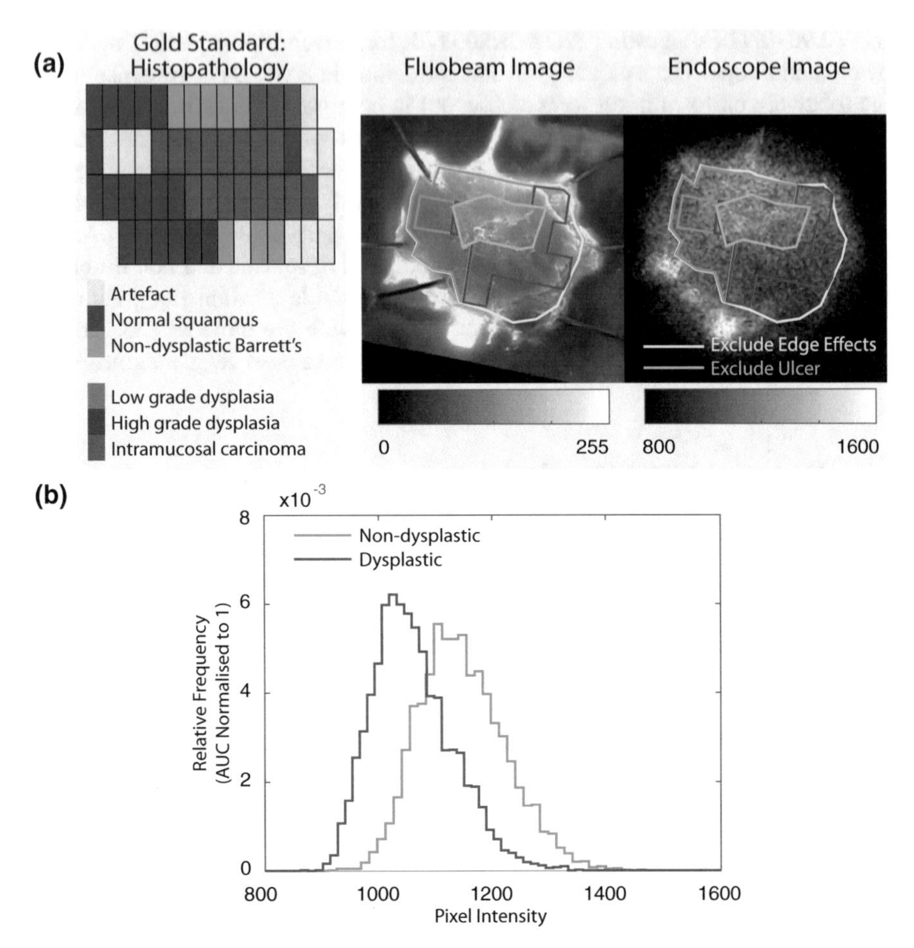

Fig. 4.10 Example co-registration with pathology from EMR B. **a** Histology grid manually coregistered to the Fluobeam image and then transferred to NIR endoscope image (since the Fluobeam and endoscope images were previously coregistered). The outside region was excluded to remove edge effects due to pooling of the dye between the edges of the tissue and the parafilm (yellow). An ulcer identified by the pathologist was excluded (orange). Areas labelled "artefact" in the histology grid may: contain pin holes; have no tissue present for analysis (due to the rectangular cuts); or be blurry in the scanned image. **b** To facilitate histopathological correlation, the normal squamous epithelium and oesophageal gastric/intestinal metaplasia (non-dysplastic Barrett's) were grouped together as "non-dysplastic" and any grade of neoplasia, including indefinite for dysplasia, low- and high-grade dysplasia and intramucosal cancer, were grouped together as "dysplastic". Pixel values for the endoscope image are plotted as a histogram for the largest continuous "non-dysplastic" and "dysplastic" regions

To achieve accurate coregistration of single points, punch biopsies can be taken from EMRs. These small (~2 mm diameter) discs of tissue, allow a single point pathology classification to be accurately coregistered with a fluorescence image using the clear hole left by the punch biopsy, which is visible in the fluorescence image (Sect. 4.6.1.2). However, the current approved protocol only allows two punch biopsies to be taken from each EMR, so many EMRs are needed to build a dataset large enough to perform statistical tests (Translational Barrier 3).

To gather spatial data, the challenge of coregistering pathology grids with fluorescence images remains. In future work, the process of coregistration could be improved by developing a flexible coregistration algorithm. Fiducial markers, such as spots of dye, could provide reference points in both the fluorescence image and histopathological sections. Using these markers, the algorithm could estimate the transformation of the pathology grid due to the non-uniform translation, rotation and stretching of the tissue that occurs during histopathological preparation. However, since EMR histopathology results have a direct impact on the subsequent care of the patient, changes to the histopathological preparation process face challenges in gaining approval.

While the results presented in this chapter are promising, some instrumentation challenges remain for clinical translation. In this work, ex vivo specimens were imaged on a flat surface. Optimal sensitivity in this geometry was found for the shortest WDs in the range that can be expected in vivo. Clinical endoscopy may encompass WDs spanning a range of several centimetres within a single image. To reliably identify dysplasia in these images, it is crucial to distinguish between areas where fluorescence signal is low due to lack of WGA-IR800 binding (true positive for dysplasia), and areas where fluorescence signal is low due to increased WD (false positive for dysplasia). Thus, fluorescence images must be corrected for the variable WD.

In addition, a number of developments could be included to ensure that we achieve acceptable SNR at a wider range of WDs. We could employ a more tailored light source, such that more spectral power is deposited across the excitation band of the IR800, or increase the overall illumination power of the system given that we remain significantly below the power used in current commercial endoscopes.

Application in a clinical setting in vivo also requires optimisation of the dye concentration to be used with our system. The sensitivity assays performed here used solutions of the dye in well plates. An important next step would be to calibrate how the sprayed concentration of WGA-IR800 relates to the final bound concentration observed on the tissue following application and washing with a spray catheter. A quantitative assessment of the applied concentration of dye, and specific/non-specific binding against a known concentration of sialic acid residues, would allow us to determine the optimum spraying concentration for use in future clinical trials.

Despite these outstanding challenges, the technique has several attractive advantages. WGA-IR800 can be applied topically using a spray catheter; requires only 10 min incubation time, minimising the disruption to the normal clinical workflow; and is easily displaced by washing with an excess of glucosamine following imaging (Translational Characteristic 1). Our endoscope is robust and compact; provides a coregistered white light image (Translational Characteristic 6); is based around a CE marked accessory channel endoscope, facilitating local safety approval for use in humans; is compatible with insertion through the working channel of standard of care endoscopes, enabling easy implementation during standard of care procedures (Translational Characteristic 2); and can be articulated using the familiar standard of care endoscope (Translational Characteristic 3). Fluorescence images are easy to interpret (Translational Characteristic 4), with dysplasia simply represented as low intensity, in contrast to a pattern based OIB, where the interpretation may require complex feature classification (e.g. mucosal and vascular patterns in NBI).

The work presented in this chapter represents progress in translating OMI with WGA-IR800 through Domain 1 and the beginning of Domain 2 of the OIB Roadmap (Fig. 1.3). With the majority of the initial validation completed using ex vivo tissue, and the device prepared for application to be locally approved for use in humans, the next step in the roadmap is to perform first-in-human trials.

Much of the groundwork for these trials has already been laid, but in order to ensure the quality and safety of WGA-IR800, it must be synthesised under good manufacturing practice (GMP) conditions before it can be used in humans. Unfortunately, the prohibitively high cost of synthesising a sufficient amount of WGA-IR800 under GMP conditions has thus far prevented the progression of this work to first-in-human trials. Though several avenues were explored to overcome this challenge, including some initially promising academic collaborations, work to find a solution is still ongoing, and could not be achieved within the timescale of the work presented in this thesis.

Still, this delay afforded us an opportunity to address some of the aforementioned instrumentation challenges. One approach is to use multispectral imaging. The ability to acquire spectral information has the potential to allow multiplexed fluorescence imaging of multiple targeted fluorophores. This could be used to increase detection specificity, by enabling the detection of consensus of several targeted molecular imaging probes, or to increase sensitivity by enabling the detection of several molecular imaging probes targeting multiple disease phenotypes. Furthermore, gathering spectral information allows working distance correction, by using data from reflectance detection to normalise the fluorescence signal [5, 28, 29]. Finally, multispectral imaging has the potential to allow delineation of dysplasia based on endogenous contrast alone, circumventing the aforementioned challenges with GMP synthesis of WGA-IR800.

In the next chapter, the development of multispectral endoscopy is described. The development of this technique for endoscopic detection of multiple fluorescent contrast agents, and for detection of endogenous contrast is described, the latter of which allowed us to make further progress towards clinical translation, carrying out the first-in-human trial of the multispectral endoscope.

References

1. James ML, Gambhir SS (2012) A molecular imaging primer: modalities, imaging agents, and applications. Physiol Rev 92:897–965
2. Sturm MB, Wang TD (2015) Emerging optical methods for surveillance of Barrett's oesophagus. Gut 64:1816–1823
3. Sevick-Muraca EM et al (2013) Advancing the translation of optical imaging agents for clinical imaging. Biomed Optics Express 4:160–170
4. Sturm MB et al (2013) Targeted imaging of esophageal neoplasia with a fluorescently labeled peptide: first-in-human results. Sci Transl Med 5:184ra61
5. Joshi BP et al (2016) Multimodal endoscope can quantify wide-field fluorescence detection of Barrett's neoplasia. Endoscopy 48
6. Bird-Lieberman EL et al (2012) Molecular imaging using fluorescent lectins permits rapid endoscopic identification of dysplasia in Barrett's esophagus. Nat Med 18:315–321
7. Waterhouse DJ et al (2016) Design and validation of a near-infrared fluorescence endoscope for detection of early esophageal malignancy. J Biomed Optics 21:084001
8. Habibollahi P et al (2012) Optical imaging with a cathepsin B activated probe for the enhanced detection of esophageal adenocarcinoma by dual channel fluorescent upper GI endoscopy. Theranostics 2:227–234
9. Funovics MA et al (2003) Miniaturized multichannel near infrared endoscope for mouse imaging. Molecular imaging 2:350–357
10. Realdon S et al (2015) In vivo molecular imaging of HER2 expression in a rat model of Barrett's esophagus adenocarcinoma. Dis Esophagus 28:394–403
11. Nakai Y, Shinoura S, Ahluwalia A, Tarnawski AS, Chang KJ (2012) Molecular imaging of epidermal growth factor-receptor and survivin in vivo in porcine esophageal and gastric mucosae using probe-based confocal laser-induced endomicroscopy: proof of concept. J Physiol Pharmacol 63:303–307
12. Li M et al (2010) Affinity peptide for targeted detection of dysplasia in Barrett's esophagus. Gastroenterology 139:1472–1480
13. Wong Kee Song LM et al (2011) Autofluorescence imaging. Gastrointestinal Endoscopy 73:647–650
14. von Holstein CS et al (1996) Detection of adenocarcinoma in Barrett's oesophagus by means of laser induced fluorescence. Gut 39:711–716
15. Neves AA et al (2018) Detection of early neoplasia in Barrett' s esophagus using lectin-based near-infrared imaging: an ex vivo study on human tissue. Endoscopy 50:618–625
16. Spectra Viewer | Chroma Technology Corp. Available at: https://www.chroma.com/spectra-viewer. Accessed: 22nd Aug 2018
17. LI-COR. IRDye® Infrared Dyes. Available at: https://www.licor.com/documents/eukougbp4lxjupcds7pqdjqyom13incs. Accessed: 22nd Aug 2018
18. Garcia-Allende PB et al (2013) Towards clinically translatable NIR fluorescence molecular guidance for colonoscopy. Biomed Optics Express 5:78–92
19. Tjalma JJ et al (2016) Molecular-guided endoscopy targeting vascular endothelial growth factor a for improved colorectal polyp detection. J Nucl Med (official publication, Society of Nuclear Medicine) 57:480–486
20. Sheth RA et al (2016) Pilot clinical trial of indocyanine green fluorescence-augmented colonoscopy in high risk patients. Gastroenterol Res Pract 2016:6184842
21. Glatz J et al (2014) Near-infrared fluorescence cholangiopancreatoscopy: initial clinical feasibility results. Gastrointest Endosc 79:664–668
22. Sato K, Nagaya T, Choyke PL, Kobayashi H (2015) Near infrared photoimmunotherapy in the treatment of pleural disseminated NSCLC: preclinical experience. Theranostics 5:698
23. ICNIRP (2013) ICNIRP guidelines on limits of exposure to incoherent visible and infrared radiation. Health Physics 71:804–819
24. Directive 2006/25/EC of the European Parliament and of the Council (2006)

25. Electron-Multiplying (EM) Gain. Available at: http://www.qimaging.com/resources/pdfs/emccd_technote.pdf
26. Pelli DG, Bex P (2013) Measuring contrast sensitivity. Vision Res 90:10–14
27. Kaye PV et al (2009) Barrett's dysplasia and the Vienna classification: reproducibility, prediction of progression and impact of consensus reporting and p53 immunohistochemistry. Histopathology 54:699–712
28. Joshi BP et al (2016) Multimodal video colonoscope for targeted wide-field detection of nonpolypoid colorectal neoplasia. Gastroenterology 150:1084–1086
29. Yang C, Hou V, Nelson LY, Seibel EJ (2013) Color-matched and fluorescence-labeled esophagus phantom and its applications. J Biomed Optics 18:26020

Chapter 5
Flexible Endoscopy: Multispectral Imaging

5.1 Multispectral Fluorescence Imaging of Targeted Fluorescent Molecules

Multispectral imaging (MSI) enables both spatial (x, y) and spectral (wavelength, λ) information to be recorded. This allows delineation of fluorophores applied during molecular endoscopy [1] based on their spectral properties, rather than a single intensity reading, allowing multiplexed fluorescence imaging of multiple targeted fluorophores [2, 3]. This could be used to increase detection specificity, by enabling detection of consensus of several targeted molecular imaging probes, or to increase sensitivity to a range of different pathologies by allowing detection of several molecular imaging probes targeting multiple disease phenotypes. It also has the potential to allow control for confounding variables in molecular imaging, for example, by applying an untargeted control dye to account for non-specific binding or non-uniformity of dye application. Furthermore, gathering spectral information allows working distance correction, by using data from diffuse reflectance detection to normalise the fluorescence signal [4–6]. Given this potential, a multispectral endoscope was developed to achieve multispectral fluorescence imaging (MFI) for application to optical molecular imaging of WGA (Chap. 4).

5.2 Multispectral Imaging of Endogenous Contrast

In addition to its potential for MFI, MSI also has the potential to image endogenous contrast. When it interacts with tissue, light is scattered by endogenous structures such as organelles, cell membranes and cell nuclei and it is absorbed by endogenous chromophores such as melanin and haemoglobin (Fig. 1.1). The overall propagation of light through tissue depends on the sizes, shapes, distribution and refractive indices of these scatters and absorbers. Given the complexity of biological tissue, adequately modelling this propagation is complex.

D. J. Waterhouse, *Novel Optical Endoscopes for Early Cancer Diagnosis and Therapy*, Springer Theses, https://doi.org/10.1007/978-3-030-21481-4_5

To simplify the model, a common strategy is to treat tissue as a bulk medium with a uniform distribution of scatterers and absorbers. Light propagation is then characterised by a diffusion equation with two wavelength dependent bulk parameters, the absorption coefficient $\mu_a(\lambda)$ and the reduced scattering coefficient $\mu_s^{'}(\lambda)$.

Propagation of broad diffuse illumination through epithelial tissue, as applies to wide field imaging in Barrett's oesophagus, is not well characterised, but models describing the propagation of light from a surface point source have been extensively developed for use in diffuse reflectance spectroscopy (DRS). Furthermore, DRS probes have been deployed in vivo allowing $\mu_a(\lambda)$ and $\mu_s^{'}(\lambda)$ to be derived for oesophageal tissue ($\mu_a(500\ \text{nm}) = 0.3–0.7\ \text{mm}^{-1}$ and $\mu_s^{'}(500\ \text{nm}) = 1.5–2.1\ \text{mm}^{-1}$ [7–11]).

At a given wavelength, the values of these coefficients can be used to calculate an optical penetration depth, which is ~1 mm for green visible light (550 nm), ~2 mm for red visible light (700 nm) and ~3 mm for NIR light (800 nm) [10], sufficiently deep to image the epithelium (~0–500 μm) where dysplastic cells are found. These are consistent with the reported contrast mechanism of narrow band imaging, which uses green light to image the superficial submucosa, approximately 1–2 mm below the surface, allowing visualisation of submucosal vessels in Barrett's surveillance.

Disease-related biochemical changes in the epithelial layer might alter the distribution and abundance of absorbers and scatterers, resulting in subtle changes in the diffuse reflectance spectrum, which might then be used to reveal the underlying pathology. The potential of this spectral data has been demonstrated in a wide range of potential applications in biomedical imaging [12, 13].

MSI can extend the acquisition of spectral image data beyond the current clinically implemented endoscopic methods of autofluorescence imaging and dye-based or virtual chromoendoscopy [14]. In combination with data analysis using spectral unmixing algorithms, MSI has been used to visualise the vascular pattern and the oxygenation status of blood [3, 15–20]; to improve detection of gastric [21] and colorectal lesions [22–24]; to identify residual tumour [25]; and to perform tissue segmentation [26, 27].

To capitalise on the potential of this rich endogenous contrast for delineation of dysplasia, in parallel to our work on MFI, a multispectral endoscope for reflectance imaging was developed. We aimed to carry out a first-in-human clinical trial of this device to gain experience operating the device in a real clinical setting, including training of endoscopists in deploying the device through the accessory channel of a standard of care endoscope, to identify improvements required for future in vivo work, and to investigate the potential of multispectral reflectance imaging for delineation of dysplasia in surveillance of Barrett's oesophagus.

5.3 Spectrally Resolved Detector Arrays (SRDAs)

The majority of spectral imaging devices fall into two categories: amplitude division, where the light beam is divided into two new beams; and field division, where the light is filtered or divided based on its position in the beam [28]. Previously reported spectral endoscopy systems use amplitude division, including multiple bandpass filters [23, 29], tuneable filters [22, 30], laser lines [31–33], or detectors dedicated to separate spectral bands [31, 32]. Amplitude division requires the use of multiple expensive optical components, making these systems both bulky, costly and more susceptible to mis-alignment in a clinical environment (Translational Characteristic 2). Furthermore, the sequential acquisition necessary with many of these systems results in slow acquisition rates, unsuitable for real-time clinical imaging.

Spectrally resolved detector arrays (SRDAs) divide the light field using spectral filters deposited directly onto the imaging detector in a mosaic pattern (Fig. 3.5). The deposition of filters directly onto the sensor results in a compact and robust device, much more suitable to clinical translation than beam splitting alternatives. Furthermore, in contrast to devices using multiple lenses, mirrors, gratings or detectors, SRDAs require no more alignment than would be needed to implement a monochrome sensor.

Although the mosaic of filters introduces an inherent trade-off between spectral and spatial resolution, since the resolution of fibrescope based imaging is limited by the size of individual fibrelets rather than by the sensor resolution, SRDAs can be implemented in fibrescopic imaging without reducing resolution (Chap. 3). These factors, along with their commercial availability, low cost and acquisition speed, underline SRDAs as a highly suitable solution for multispectral endoscopic imaging (Translational Characteristic 2) [3, 34], motivating the development of an SRDA-based multispectral endoscope.

For the remainder of this chapter, the preliminary work on the development of SRDA-based multispectral endoscopy is described. The design, characterisation and preliminary validation of a clinically translatable SRDA-based multispectral endoscope with two alternative optical configurations, one for fluorescence imaging of fluorescent contrast agents, and one for reflectance imaging of endogenous tissue contrast (Sect. 5.4) is presented. The feasibility of using this device to perform multispectral fluorescence imaging (MFI) is assessed in Sect. 5.5 using ex vivo models. Following this, multispectral reflectance imaging is investigated in a first-in-human trial in Sect. 5.6.

5.4 Materials and Methods

5.4.1 Fluorescent Contrast Agents

The main disadvantage of SRDAs is their lower sensitivity compared to EMCCDs such as that used in Chap. 4, so the fluorescence signal is particularly important. We therefore decided to use AlexaFluor-647 (AF647, Thermo Fisher Scientific, USA) for this work, as it exhibits significantly brighter fluorescence than IR800. Furthermore, high powered LED sources are more readily available in the far-red than in the NIR, allowing additional power to be deposited across the excitation peak of AF647, further increasing the fluorescence signal. The AlexaFluor series of dyes are also available with fluorescence emission peaks at a range of closely spaced wavelengths, so multiplexed imaging of two similar fluorophores can be assessed. Additionally, AF647 is commercially available as a WGA conjugate, WGA-AF647 (Thermo Fisher Scientific, USA). It was hoped that this would improve our chances of achieving translation to first-in-human trials.

5.4.2 Endoscope Design

A multispectral endoscope based around the PolyScope accessory channel endoscope (Sect. 3.2) and a compact SRDA (CMS-V, SILIOS, France) was designed as shown in Fig. 5.2. The SRDA consists of 9 spectral filters (8 narrow bands; average FWHM 30 nm; centre wavelengths 553, 587, 629, 665, 714, 749, 791, 829 nm; 1 broad band; 500–850 nm), deposited as a 3 $\times$ 3 super-pixel across a CMOS sensor (NIR Ruby sensor, UI1242LE-NIR, IDS, square pixel size 5.3 μm). Due to the compact SRDA, the system is much smaller than the bimodal endoscope (Sect. 4.4.2) as shown in Fig. 5.1. The optics are securely housed inside the same portable light tight enclosure described in Sect. 4.4.2.

The device has two alternative optical configurations.

- For fluorescence imaging (Fig. 5.2a), illumination is provided by a narrow band ultra-high power LED (Fig. 5.3) (UHP-T-LED-635-EP, Prizmatix, Israel) coupled into the PolyScope illumination channel using an achromatic doublet lens (AC254-030-A, Thorlabs, Germany) housed inside a custom coupler with a smooth bore for the PolyScope illumination fibre tip. An objective lens (NA = 0.5, UPLFLN20x,

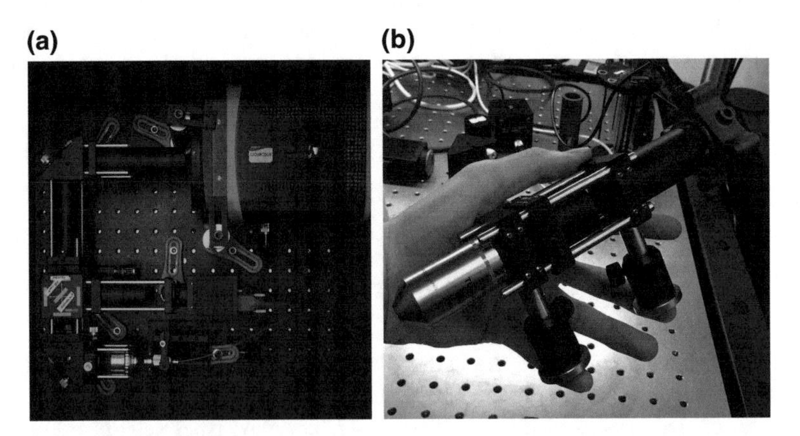

Fig. 5.1 SRDA-based endoscope compared to bimodal NIR fluorescence and WL reflectance endoscope. The SRDA-based multispectral endoscope (**b**) is more compact than the bimodal NIR fluorescence and WL reflectance endoscope (**a**) described in Chap. 4 as it requires fewer optical components and utilises a single compact SRDA

Olympus, Japan) and an achromatic doublet lens ($f = 100$ mm, ACA254-100-A, Thorlabs, Germany), focus light from the 10,000-fibrelet bundle onto a compact SRDA (CMS-V, SILIOS, France) allowing multispectral fluorescence imaging.

- For reflectance imaging (Fig. 5.2b), the illumination is replaced with a broadband UHP-LED (Fig. 5.3) (T7359, Prizmatix, Israel). A 90:10 beam splitter (BSN10R, Thorlabs, Germany) relays 10% of the light, through an achromatic triplet lens ($f = 40$ mm, TRH254-040-A-ML, Thorlabs, Germany), into a 1000 μm fibre (M35L01, Thorlabs, Germany) coupled to a spectrometer (AvaSpec-ULS2048, Avantes, Netherlands). This was used to acquire a single gold standard reflectance measurement of the entire image area during clinical imaging.

5.4.3 Image Acquisition and Image Corrections

For the non-clinical fluorescence system (Fig. 5.2a), images were captured in uEye Cockpit (IDS, Germany) and saved as 8-bit BMP files. Videos were captured in uEye Cockpit (IDS, Germany) and saved as AVI files. Data analysis was carried out using Matlab® (MathWorks, USA). For the clinical reflectance system (Fig. 5.2b), a LabVIEW (National Instruments, USA) Visual Interface (VI) was developed to capture and display the images and spectra for the endoscopist to view them in real-time. The raw images captured by the SRDA were demosaicked and decombed using the interpolation methods outlined in Sect. 3.3.3. For the clinical system, a false colour RGB image was generated by assigning the narrow bands centred at 629, 587 and 553 nm to RGB channels respectively.

5.4.4 Spectral Unmixing

For each spatial (x, y) position in a multispectral image, spectral (wavelength, λ) information is acquired. Each raw spectrum is a sum of reflection and fluorescence from any fluorophores present, those exogenously applied or those endogenous to tissue. To make use of the spectral information, "unmixing" the individual spectra, or "endmembers", from each of these contributions is of interest. In MFI, these endmembers are the reflected illumination spectrum and the fluorescence emission spectrum of each fluorophore. In reflectance imaging of endogenous tissue contrast, the endmembers might be the spectra of individual disease phenotypes or the individual reflection spectra of diagnostically relevant biological molecules. To extract the abundance of each endmember from the raw spectrum, it is necessary to perform spectral unmixing. In this work non-linear least squares unmixing was used to fit linear sums of modelled endmembers to the raw spectra captured by the endoscope.

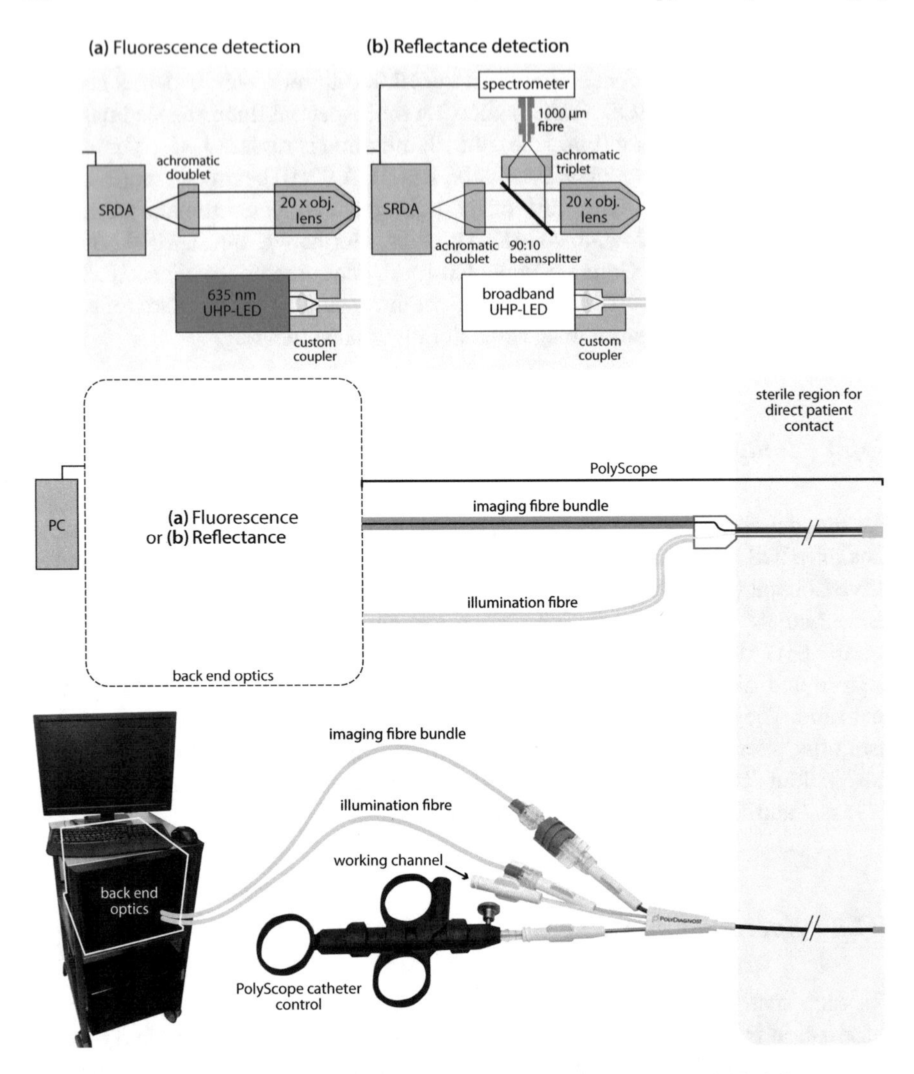

Fig. 5.2 SRDA-based multispectral endoscope. The system is based around the PolyScope accessory channel endoscope (PolyDiagnost, Germany). The device has two alternative optical configurations. **a** For fluorescence imaging, illumination is provided by a narrow band ultra-high power LED (UHP-T-LED-635-EP, Prizmatix, Israel) coupled into the PolyScope illumination channel using a custom coupler. An objective lens (NA = 0.5, UPLFLN20x, Olympus, Japan) and an achromatic doublet lens ($f = 100$ mm, ACA254-100-A, Thorlabs, Germany), focus light from the 10,000-fibrelet bundle onto a compact SRDA (CMS-V, SILIOS, France) allowing multispectral fluorescence imaging. **b** For reflectance imaging, the illumination is replaced with a broadband UHP-LED (T7359, Prizmatix, Israel). A 90:10 beam splitter (BSN10R, Thorlabs, Germany) relays 10% of the light, through an achromatic triplet lens ($f = 40$ mm, TRH254-040-A-ML, Thorlabs, Germany), into a 1000 μm fibre (M35L01, Thorlabs, Germany) coupled to a spectrometer (AvaSpec-ULS2048, Avantes, Netherlands), to acquire a single gold standard reflectance measurement of the entire image field. The pink area of the PolyScope catheter must remain sterile once unpacked, as this part is in direct contact with the patient during endoscopy

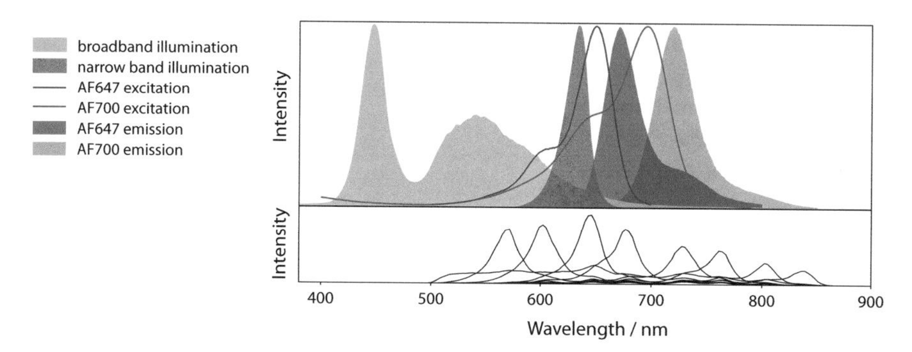

Fig. 5.3 Spectra of the components and contrast agents used for multispectral imaging. The spectra of the light sources used for illumination in MFI and multispectral reflectance imaging, narrow band UHP-LED and broadband UHP-LED respectively (data from Prizmatix). Excitation and emission spectra of AF647 and AF700 are shown (data from ThermoFisher [35]). The spectral response of the multispectral endoscope is shown below for reference (data from Sect. 5.5.1.3)

5.5 Multispectral Fluorescence Imaging (MFI)

After the SRDA-based fluorescence endoscope was designed and built, characterisation was carried out with respect to the safety of the system, by quantifying its illumination power (Sect. 5.5.1.1) and with respect to its intended operation, by quantifying its resolution (Sect. 5.5.1.2), FOV (Sect. 5.5.1.2) and spectral response (Sect. 5.5.1.3). Following this, the device was validated for MFI, first using a dilution series of AF647 (Sect. 5.5.1.4), and then using fluorescent phantoms containing AF647 and AF700 placed in a whole ex vivo pig oesophagus, simulating clinical endoscopy (Sect. 5.5.1.5).

5.5.1 Methods

5.5.1.1 Power

As previously discussed in Sect. 4.5.1.2, defining a maximum permissible exposure for in vivo optical imaging is difficult due to a lack of published safety standards. Thus, to establish a safe maximum power for the multispectral endoscope, its maximum broadband power was compared to that of a clinically approved standard of care endoscope as described in Sect. 4.5.1.2.

5.5.1.2 Resolution and FOV

The resolution of the device was determined as described in Sect. 3.3.5.1. The FOV is determined by the PolyScope accessory channel endoscope, so the FOV of the multispectral endoscope is the same as that measured for the bimodal endoscope in Sect. 4.5.1.1.

5.5.1.3 Spectral Response

In order to model endmembers for spectral unmixing, "ground truth" spectra of interest must be propagated through the spectral response of the multispectral endoscope, so it is important that this is characterised. To characterise the spectral response of the multispectral endoscope, a broadband halogen light source (OSL2BIR, Thorlabs, Germany) was fibre coupled into a monochromator (CM110, AG1200-00500-303 grating, blaze wavelength 500 nm, grating density 1200 mm^{-1}, Spectral Products, USA), which was used to increment the wavelength between 450–900 nm (with a step size of 3 nm and a FWHM of ~2 nm). The resulting narrow bands of illumination were directed into the detection pathway of the multispectral endoscope (Fig. 5.4a). At each wavelength step, the SRDA captured 10 images.

Subsequently, the detection arm of the multispectral endoscope was replaced with a spectrometer (AvaSpec-ULS2048, Avantes, Netherlands) (Fig. 5.4b), and the monochromator was scanned again; an illumination spectrum was acquired at each scan wavelength. For each scan wavelength, a dark image was subtracted from each of the 10 multispectral images. The mean pixel intensity across the ten images was

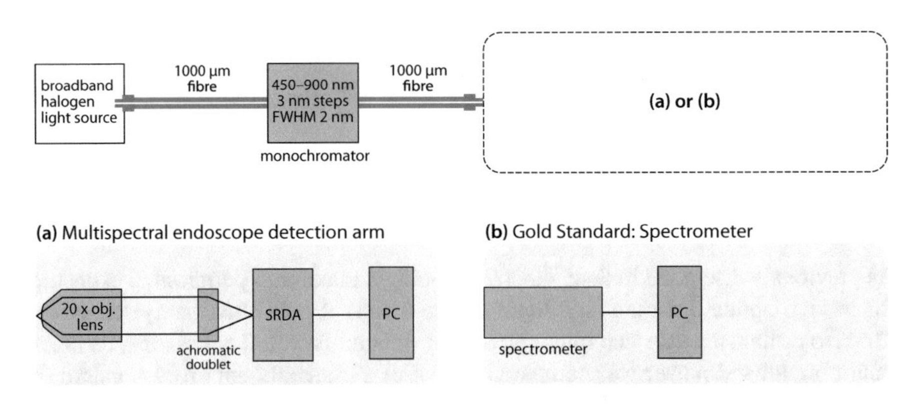

Fig. 5.4 Spectral response characterisation equipment. A broadband halogen light source (OSL2BIR, Thorlabs, Germany) was fibre coupled into a monochromator (CM110 Compact 1/8 Monochromator, AG1200-00500-303 grating, blaze wavelength 500 nm, grating density 1200 mm^{-1}, Spectral Products USA), which was used to increment the wavelength between 450–900 nm (with a step size of 3 nm and a FWHM of ~2 nm). This was directed into **a** the multispectral endoscope and **b** a spectrometer (AvaSpec-ULS2048, Avantes, Netherlands)

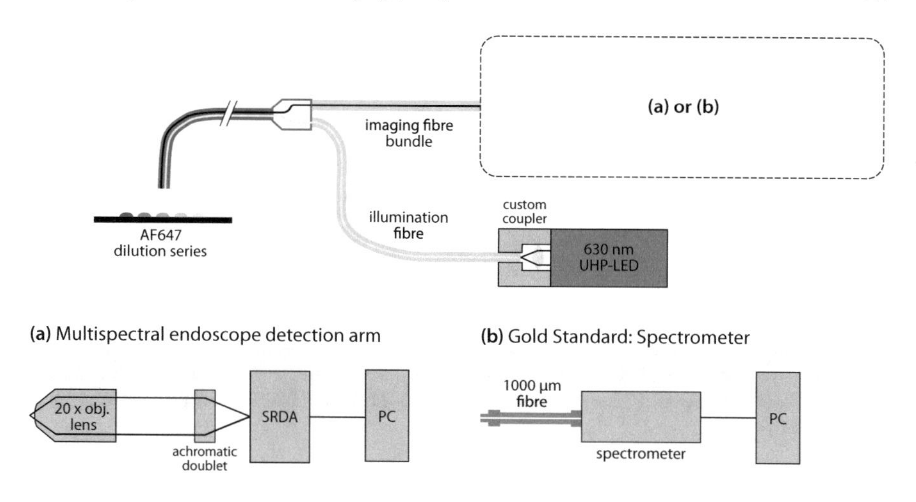

Fig. 5.5 Fluorescence detection characterisation equipment. **a** The detection arm of the multispectral fluorescence endoscope is used to capture images of a 2:1 dilution series of Alexa-Fluor-647 in a well plate (ibidi, Germany). **b** Immediately following this, the multispectral endoscope detection pathway is replaced by a spectrometer (AvaSpec-ULS2048, Avantes, Netherlands) to measure the AF647 spectra. These spectra can be used to determine the ground truth concentration

calculated and this value was divided by the peak intensity of the illumination spectrum to yield the final spectral response. This calculation was performed at each scan wavelength to yield the spectral response curves.

5.5.1.4 Multispectral Fluorescence Imaging of AF647 in Solution

To validate the performance of the multispectral fluorescence endoscope in MFI, a dilution series of AF647 (30 μL of a 1:2 dilution from 1–0.0039 mg/mL) was prepared in a clear well plate (ibidi, Germany). Images of the dilution series were captured using the multispectral fluorescence endoscope (Fig. 5.5a). As a gold standard comparison, the detection pathway of our endoscope was replaced with a spectrometer (AvaSpec-ULS2048, Avantes, Netherlands) to measure the 'ground truth' spectrum from each concentration of dye (Fig. 5.5b). This was repeated 5 times with removal and replacement of the sample. The average spectrum was calculated from the 5 repeats for each concentration.

5.5.1.5 Multispectral Fluorescence Imaging of Multiple Fluorophores in Ex Vivo Pig Oesophagus

After testing the ability of the multispectral fluorescence endoscope to detect a single fluorophore by unmixing fluorescence and reflectance endmembers in images of an

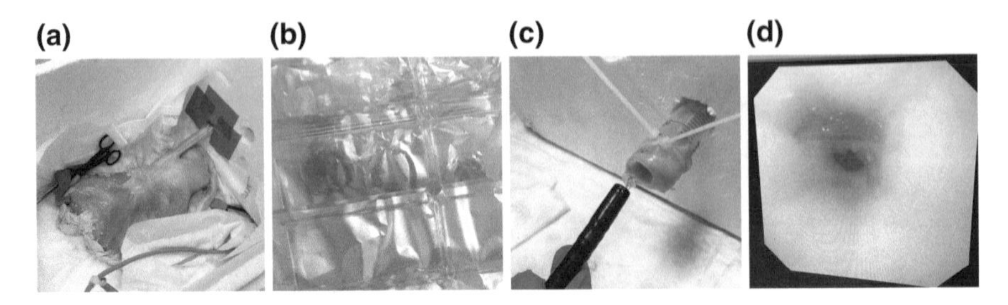

Fig. 5.6 Procedure for preparing ex vivo porcine oesophagus for endoscopic imaging. **a** Whole ex vivo porcine oesophagus. **b** Agarose tissue mimicking phantoms containing PBS, AF647 and AF700 were prepared. **c** The agarose phantoms were inserted inside the ex vivo porcine oesophagus using endoscopic forceps. **d** The phantoms adhered to the luminal surface of the oesophagus by surface tension

AF647 dilution series, the ability of the endoscope to image multiple fluorophores in a more realistic endoscopic scenario was tested. A clinical endoscopic setting was simulated using a whole ex vivo porcine oesophagus (Fig. 5.6a) (Med Meat Supplies, UK). For fluorescence targets, agarose tissue mimicking phantoms containing AlexaFluor 647 (AF647), AlexaFluor 700 (AF700) and PBS as a control were prepared (Fig. 5.6b).

Briefly, the phantoms were prepared by mixing equal parts of a 6.0% agar solution with 160 μM fluorescent dye dilutions in PBS. The liquid agarose was pipetted onto glass slides to cool and set to form ~5 mm diameter, ~3 mm thick droplets. A trained endoscopist (Massimiliano di Pietro) used endoscopic forceps via the working channel of a clinical endoscope (Fig. 5.6c) (GIF-H260, Olympus, Japan), to carefully place these phantoms inside the ex vivo porcine oesophagus, where they adhered to the luminal surface by surface tension (Fig. 5.6d).

Next, the multispectral endoscope was taped to the side of the clinical endoscope since the working channel of the endoscope in use was only 2.8 mm, too small for the diameter of the 3 mm PolyScope catheter. Video was recorded at ~5 frames per second as the endoscope and PolyScope were slowly moved into the oesophagus and then slowly withdrawn.

5.5.2 *Results*

5.5.2.1 Power

To establish a safe maximum power for our multispectral endoscope, its maximum broadband power was compared to that of a clinical endoscope. The maximum power from the clinical endoscope in white light reflectance mode was measured to be 19 $\pm$ 1 mW, at a working distance of 1.0 $\pm$ 0.1 cm. The maximum powers in fluorescence

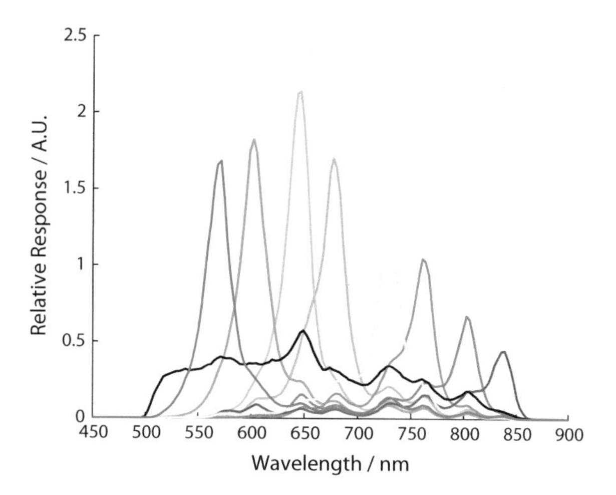

Fig. 5.7 The spectral response of the detection arm of the SRDA-based multispectral endoscope

imaging configuration and in reflectance imaging configuration were measured to be 12.9 ± 0.5 mW and 10 ± 1 mW respectively, at a working distance of 1.0 ± 0.1 cm.

5.5.2.2 Resolution and FOV

The resolution of the device is 240 ± 20 μm as determined using the methods described in Sect. 3.3.5.1. The FOV is governed by the PolyScope accessory channel endoscope, so the FOV of the multispectral endoscope is $63 \pm 1°$, as quantified for the bimodal endoscope in Sect. 4.5.1.1.

5.5.2.3 Spectral Response

The spectral response of the SRDA-based multispectral endoscope determined using the spectrometer and monochromator is shown in Fig. 5.7.

5.5.2.4 Multispectral Fluorescence Imaging of AF647 in Solution

To validate the performance of the multispectral fluorescence endoscope in MFI, an AF647 dilution series was prepared. Using a spectrometer, the 'ground truth' spectra, $S_{AF647}(C, \lambda)$, of an AF647 dilution series were recorded, where C denotes the concentration (Fig. 5.8a). The reflection spectrum (~600 to 650 nm) shows a strong dependence on spatial position of the sample, indicated by the large shaded standard deviation of the 5 measurements. This is expected since reflection is non-isotropic and thus depends on the measurement geometry. The fluorescence emission,

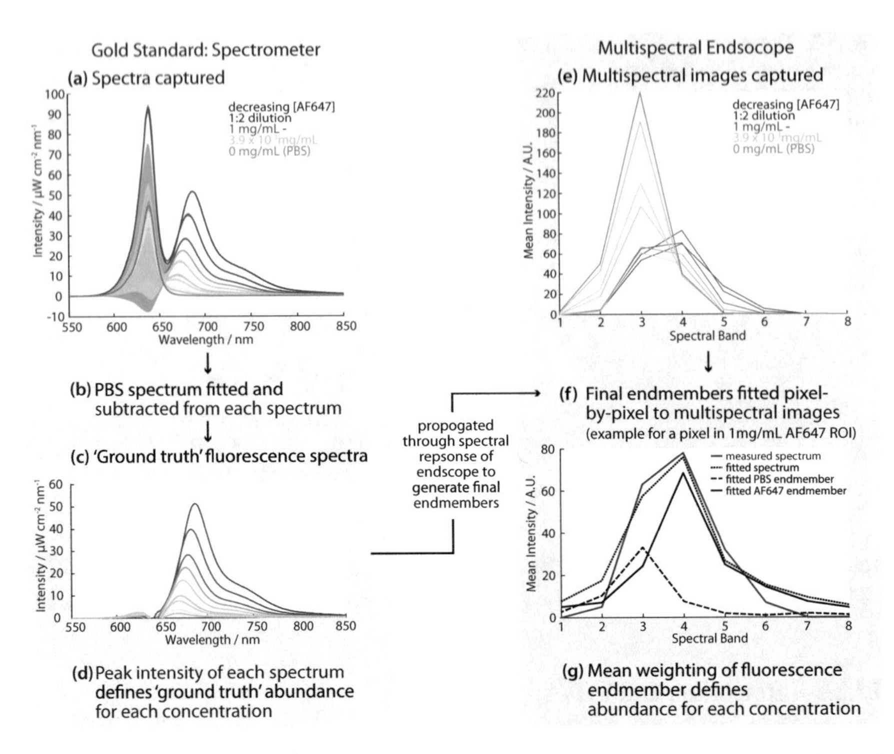

Fig. 5.8 AF647 abundance measured using the multispectral endoscope and 'ground truth' AF647 abundance measured using a spectrometer. A 1:2 dilution series of AF647 was prepared. **a** Spectra of each concentration of AF647 and of PBS were acquired using a spectrometer. The shaded region represents the standard deviation of five spectra acquired with removal and replacement of the sample. **b** The PBS spectrum was fitted to the spectrum for each concentration and subtracted to leave only the fluorescence spectra shown in (**c**). **d** The peak intensity of each of these spectra defines the 'ground truth' abundance of AF647 for each concentration. **e** Next, the multispectral endoscope was used to capture images of each concentration of AF647 and of PBS. The mean spectra for pixels in the ROI within each of the captured multispectral images are shown. The broad band is omitted for clarity. **f** A weighted sum of the fluorescence and reflectance endmembers was fitted to the image cubes on a pixel-by-pixel basis. An example is shown for a pixel in the 1 mg/mL AF647. **g** The mean weighting of the fluorescence endmember in the ROI defines the abundance of AF647 for each concentration. The multispectral endoscope has 8 narrow spectral bands; average FWHM 30 nm; centre wavelengths 553, 587, 629, 665, 714, 749, 791, 829 nm

(~650 to 750 nm), is expected to be isotropic and unaffected by specular reflections from the sample surface, and hence shows a minimal dependence on position.

The following steps were performed in order to model the reflectance and AF647 fluorescence endmembers.

(i) A least squares smoothing spline was fitted to the PBS spectrum, $S_{AF647}(0, \lambda)$, which should only include reflectance. The resulting spline determined the reflectance signal, $R(\lambda)$ (Fig. 5.8b).

(ii) This reflectance spectrum, $R(\lambda)$, was fitted (least squares) to each of the AF647 spectra, $S_{AF647}(C, \lambda)$. The resulting fitted reflectance spectra were subtracted to leave only the 'ground truth' fluorescence spectra $F_{AF647}(C, \lambda)$ (Fig. 5.8c).

(iii) The peak intensity of each of these fluorescence spectra was used to determine the 'ground truth' abundance of each concentration of dye, $A_{groundtruth}(C)$ (Fig. 5.8d).

(iv) For each of the 9 concentrations, the fluorescence spectrum, $F_{AF647}(C, \lambda)$, was propagated through the spectral response of the multispectral endoscope, $P(B, \lambda)$ (Sect. 5.5.1.3), to yield a fluorescence endmember:

$$F^{em}(C, B) = \int F_{AF647}(C, \lambda) P(B, \lambda) d\lambda \tag{5.1}$$

where B is the spectral band of our endoscope ($B = 1$–9).

(v) The resulting spectra were normalised and a mean taken across the 9 concentrations to yield a final predicted fluorescence endmember:

$$F^{em}(B) = \frac{1}{9} \sum_{C} \overline{F^{em}(C, B)} \tag{5.2}$$

where the bar denotes normalisation to area under the curve = 1.

(vi) Step (iv) was repeated for the reflectance spectrum, $R(\lambda)$, to yield the final predicted reflectance endmember, $R^{em}(B)$.

Next, images of the dilution series were captured using the multispectral fluorescence endoscope (Fig. 5.8e). A linear weighted sum of the fluorescence and reflectance endmembers was fitted to the image cubes on a pixel-by-pixel basis (Fig. 5.8f). The weighting of the fluorescence endmember defines the measured abundance of AF647 for each pixel, $A_{measured}(C, x, y)$. The mean measured abundance within the region of interest (ROI) within the multispectral image determines the final measured abundance of AF647, $A_{measured}(C)$ (Fig. 5.8f).

A strong correlation ($R^2 = 0.9816$) was found between $A_{measured}(C)$ and $A_{groundtruth}(C)$ (Fig. 5.9). The small deviation observed at the highest concentration is most likely due to poor off-centre placement of the sample. These results indicate that the multispectral endoscope is capable of accurately unmixing the highly variable background reflectance signal from the fluorescence signal of interest.

5.5.2.5 Multispectral Fluorescence Imaging of Multiple Fluorophores in Ex Vivo Pig Oesophagus

Having demonstrated the multispectral endoscope is able to accurately image AF647, unmixing two endmembers, one fluorescence and one reflectance, the capability of the endoscope to unmix three endmembers, one reflectance and two fluorescence,

AF647 and AF700, was investigated. A whole ex vivo porcine oesophagus was used to simulate clinical endoscopy.

Predicted AF647, AF700 and reflectance endmembers, $E(B)$, were calculated by propagating the data book spectra of AF647, AF700 and the 635 nm LED, $S^D(\lambda)$, through the spectral response of our multispectral endoscope, $P(B, \lambda)$ (Sect. 5.5.1.3). For example:

$$E_{AF647}(B) = \int S^D_{AF647}(\lambda) P(B, \lambda) d\lambda \tag{5.3}$$

where B is the spectral band of our endoscope ($B = 1$–9). These were normalised to area under the curve = 1.

Each frame of the resulting video was spectrally unmixed by fitting a weighted sum of these endmembers to the 9-point spectrum at each fibrelet centre pixel following demosaicking but prior to interpolation. The weighting of each endmember, AF647, AF700 and reflectance, defines the measured abundance of AF647, AF700 and reflectance respectively. It was demonstrated that the endoscope was capable of detecting both AF647 and AF700 with signal-to-background ratios of 310 ± 90 and $(2.2 \pm 0.9) \times 10^4$ respectively (Fig. 5.10).

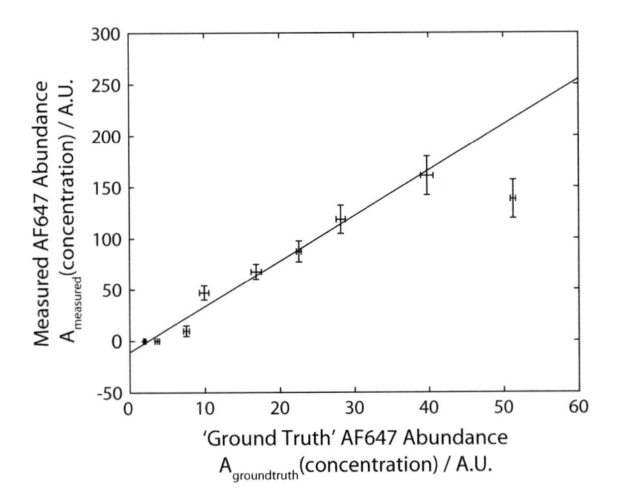

Fig. 5.9 Correlation of AF647 abundance measured using the multispectral endoscope with 'ground truth' AF647 abundance measured using a spectrometer. The measured AF647 abundance is derived by fitting endmembers to the images captured with the SRDA-based multispectral endoscope. The 'ground truth' abundance is determined from the spectrometer measurements. The error in the 'ground truth' abundance is the standard deviation from repeat measurements. The error in the measured AF647 abundance is the standard deviation of abundance over the ROI. A fit is shown for all the data points excluding the highest concentration ($R^2 = 0.9816$)

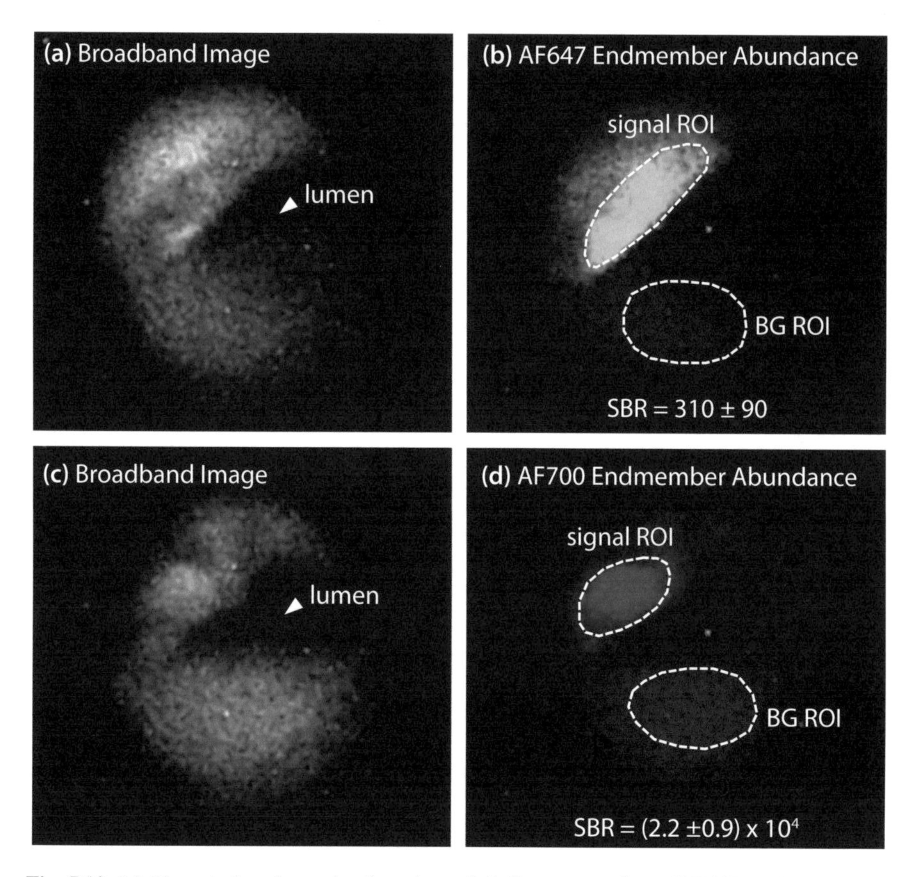

Fig. 5.10 Multispectral endoscopic detection of 2 fluorescent dyes (AF647 and AF700) in endoscopy of a whole ex vivo porcine oesophagus. **a** and **c** Broadband images of oesophageal wall with AF647 and AF700 tissue mimicking phantoms respectively. **b** and **d** False colour RGB images (Red: Broadband Image, Green: AF647 endmember abundance from unmixing, Blue: AF700 endmember abundance from unmixing) co-registered to (**a**) and (**c**) respectively. Signal and background (BG) regions of interest (ROIs) were drawn and used to calculate signal to background ratio (SBR)

5.6 Multispectral Reflectance Imaging of Endogenous Tissue Contrast

In parallel to MFI, multispectral reflectance imaging of endogenous contrast was investigated. Having developed an SRDA-based multispectral endoscope, written multispectral demosaicking and decombing algorithms, and proven the feasibility of imaging with the device in an ex vivo simulation of endoscopy, and since ex vivo tissue does not represent a good biological model of endogenous tissue contrast (Translational Barrier 4), work using multispectral reflectance endoscopy proceeded directly to in vivo trials.

Thus an in vivo single centre pilot cohort study was designed, entitled: "Prospective pilot cohort study to assess feasibility of multispectral endoscopic imaging for detection of early neoplasia in Barrett's oesophagus", or "Multispectral endoscopy (MuSE)" for short. For this trial, 20 patients will be recruited, the patient population enriched such that approximately 40–50% will have dysplasia at the study endoscopy. Imaging with the multispectral endoscope will occur as part of the subjects' clinically indicated procedure.

5.6.1 Methods

A. *Trial objectives*

The trial is designed to address two primary objectives:

(i) To assess the feasibility of imaging with the novel multispectral endoscope, ensuring the device can be used to capture images in a clinical environment and with minimal interruption to clinical workflow.
(ii) Evaluate the appearance of Barrett's oesophagus with and without dysplasia using multispectral endoscopy to build a library of potentially diagnostically useful distinguishing spectral characteristics since these are currently unknown and not easily investigated using ex vivo tissue (Translational Barrier 4).

A long-term secondary objective is also proposed, which can be addressed once a large data set of multispectral images of confirmed pathologies has been acquired and a set of imaging biomarkers to distinguish Barrett's oesophagus with and without dysplasia has been established:

(iii) Evaluate diagnostic accuracy of multispectral endoscopic image derived biomarkers for Barrett's oesophagus with dysplasia.

B. *Trial design*

Each procedure is performed by a single endoscopist.

(i) After local anaesthesia or conscious sedation (Midazolam ± Fentalyl) the endoscopist intubates the patient with a high resolution white light therapeutic endoscope (GIF-H290Z, Olympus, Japan). The use of the therapeutic scope is required given the larger accessory channel (3.7 mm diameter), which is compatible with the insertion of the PolyScope catheter (3.0 mm diameter). The endoscopist thoroughly cleans the oesophageal mucosa using standard cleansing agents. The endoscopist thoroughly inspects the mucosal surface of the oesophagus. Cautery marks are placed around suspicious lesion(s) and around one region of Barrett's oesophagus as a control (Fig. 5.11a).
(ii) The endoscopist inserts the multispectral endoscope through the working channel and uses this to inspect the Barrett's oesophagus segment with particular attention to the regions outlined in step (i) (Fig. 5.11b).

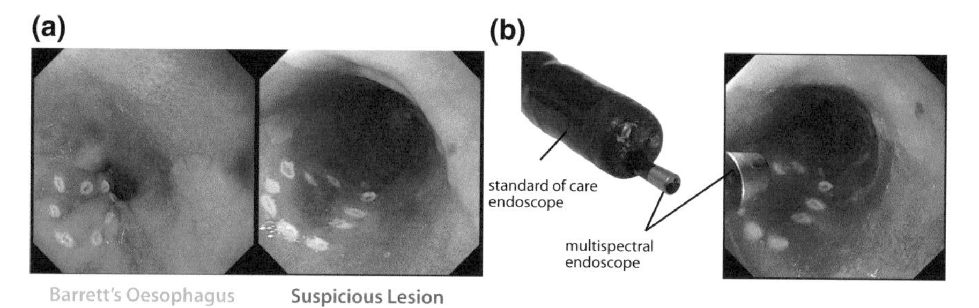

Fig. 5.11 Procedure for in vivo clinical trial of multispectral endoscopy (MuSE). **a** Suspicious lesion(s) are identified and marked using standard of care HD-WLE. **b** The multispectral endoscope is inserted via the working channel of the standard of care endoscope

(iii) The endoscopist switches back to conventional imaging. The endoscopist takes endoscopic mucosal resections from each of the suspicious lesion(s). Random biopsies are taken according to the Seattle protocol. The random biopsies are omitted if they are considered unnecessary for more advanced disease (e.g. clearly visible intramucosal cancer which needs endoscopic resection).

Pathological assessment of biopsies is performed according to the revised Vienna classification [36]. The HD-WLE video stream is recorded using a recording unit (SMP300, Extron, USA).

5.6.2 *Results (MuSE 01)*

The trial was approved on 2nd March 2018. The first patient was recruited and the first trial performed on 30th May 2018 (MuSE 01). Approximately one hour prior to the procedure, the PolyScope 10,000 fibre imaging bundle was threaded into a sterile 185 cm PolyScope catheter (PD-PS-0144p, PolyDiagnost, Germany), which was handled with sterile gloves to ensure the region intended to be in direct patient contact remained sterile (Fig. 5.2). The system was aligned, the catheter illumination fibre was connected to the light source via the custom coupler, and the power was checked to be 10 ± 1 mW using the method described in Sect. 4.5.1.2. Then, the standard of care endoscopy began with the standard of care endoscope (Lucera GIF-2T240, Olympus, Japan).

Following inspection of the whole Barrett's segment with HD-WLE, one suspicious lesion was identified. Cautery marks were placed around this suspicious lesion and around one region of Barrett's oesophagus as a control (Fig. 5.11a). Next, the multispectral endoscope was threaded into the working channel of the standard of care endoscope. Data was captured for ~12 min at ~1 fps, resulting in 842 SRDA images and 842 gold standard spectra. Finally, the multispectral endoscope was withdrawn to allow standard of care endoscopy to continue.

At this time of this procedure, the suspicious lesion was not resected due to risk of excessive bleeding from the patient. On 8th June 2018, the patient returned and the suspicious lesion was resected using endoscopic mucosal resection (EMR). The EMR specimen was sent to histopathology for diagnosis. A pathology grid was generated as described in Sect. 4.6.1.3. Histopathology confirmed the presence of moderately differentiated intramucosal adenocarcinoma (Fig. 5.12).

Analysis of the multispectral endoscopy data was carried out in Matlab® (MathWorks, USA). The raw images captured by the SRDA were demosaicked and decombed using the interpolation methods outlined in Sect. 3.3.3. A false colour multispectral endoscopy video was generated by assigning the narrow bands centred at 629, 587 and 553 nm to RGB channels respectively.

This multispectral endoscopy video and the HD-WLE video were synchronised and placed side by side for viewing (Fig. 5.13). In much of the video, the endoscopist is articulating the standard of care endoscope to bring the marked regions into the field of view of the multispectral endoscope. The video was used to identify times where this was achieved. A total of 17 images and 25 gold standard spectra, from 2 short video segments viewing the Barrett's tissue and 3 short video segments viewing the cancer tissue, were selected.

In the false colour multispectral images, the cautery marks surrounding the regions of interest appear as bright white patches. These were used to guide the drawing of regions of interest (ROIs) in the images (Fig. 5.13b). Inside each ROI, the mean across each spectral band image was calculated to yield a 9-band image spectrum. These were normalised to AUC = 1. The mean of the 9-band image spectra for Barrett's ROIs (n = 9) and cancer ROIs (n = 8) are plotted in Fig. 5.14a. These spectra show encouraging differences between the two disease types, particularly in bands 1–4 (average FWHM 30 nm; centre wavelengths 553, 587, 629, 665 nm).

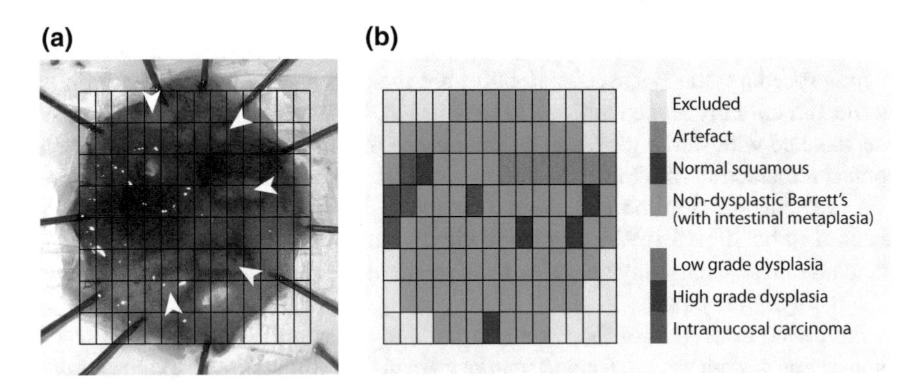

Fig. 5.12 EMR specimen from MuSE 01. **a** White light image of the EMR specimen from MuSE 01. The location of two punch biopsies are visible in row 3 column 5 and row 7 column 9. Cautery marks are also visible (white arrows). **b** Histopathology grid manually coregistered to the white light image. This confirms the presence of moderately differentiated intramucosal carcinoma in the sample

The gold standard spectra were background subtracted and normalised to AUC = 1. The mean of these spectra for Barrett's (n = 9) and cancer (n = 16) are plotted in Fig. 5.14b. These spectra show encouraging differences between the two disease types in more detail in the region from 520–650 nm. To allow comparison to the 9-band multispectral image spectra, these background subtracted, gold standard spectra, $S^g(B)$, were propagated through the spectral response of our endoscope, $P(B, \lambda)$ (Sect. 5.5.1.3) to yield modelled 9-band image spectra:

$$S^g(B) = \int S^g(\lambda)P(B, \lambda)d\lambda \tag{5.4}$$

These were normalised to AUC = 1. The mean over the modelled 9-band image spectra for Barrett's and cancer are plotted in Fig. 5.14c. These compare favourably to

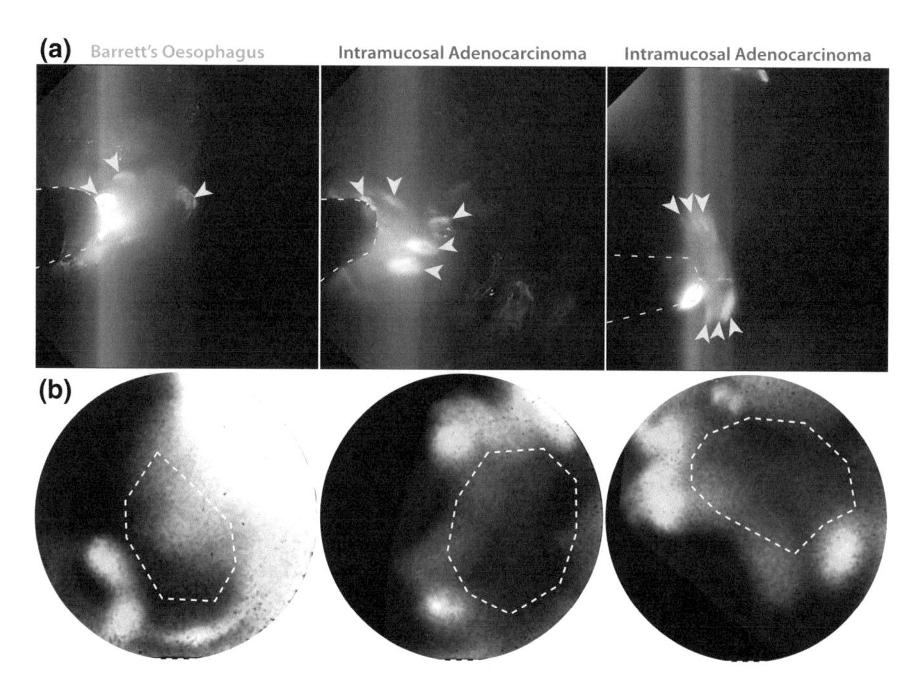

Fig. 5.13 Synchronised HD-WLE and multispectral images from the first-in-human trial of the multispectral endoscope (MuSE 01). **a** High definition white light endoscopy (HD-WLE) images from MuSE 01. Images are from three of the selected video segments. From left to right: Barrett's oesophagus, intramucosal carcinoma, intramucosal carcinoma. The cautery marks delineating the regions can be seen as bright white patches in these images (yellow arrows). The yellow dashed lines indicate the position of the multispectral endoscope to aid visualisation. The images are dark as the HD-WLE light source is switched off for multispectral imaging. **b** Multispectral images corresponding to the three HD-WLE images. These false colour RGB images are generated by assigning the narrow bands centred at 629, 587 and 553 nm to RGB channels respectively. The cautery marks are visible as bright white patches. Regions of interest (ROIs) are shown (white dashed lines)

the measured 9-band image spectra from the SRDA (Fig. 5.14a). The small standard deviation of both image spectra and gold standard spectra, shown shaded in Fig. 5.14, suggest high repeatability of the technique.

5.7 Discussion and Conclusions

To address the clinical challenge of Barrett's surveillance, an SRDA-based multispectral endoscope capable of acquisition of 9-band multispectral images in vivo was developed. Two alternative optical configurations of the device were tested. The first, designed for multispectral fluorescence imaging, was demonstrated to be capable of accurately detecting fluorescent contrast agents both in well plates and in a

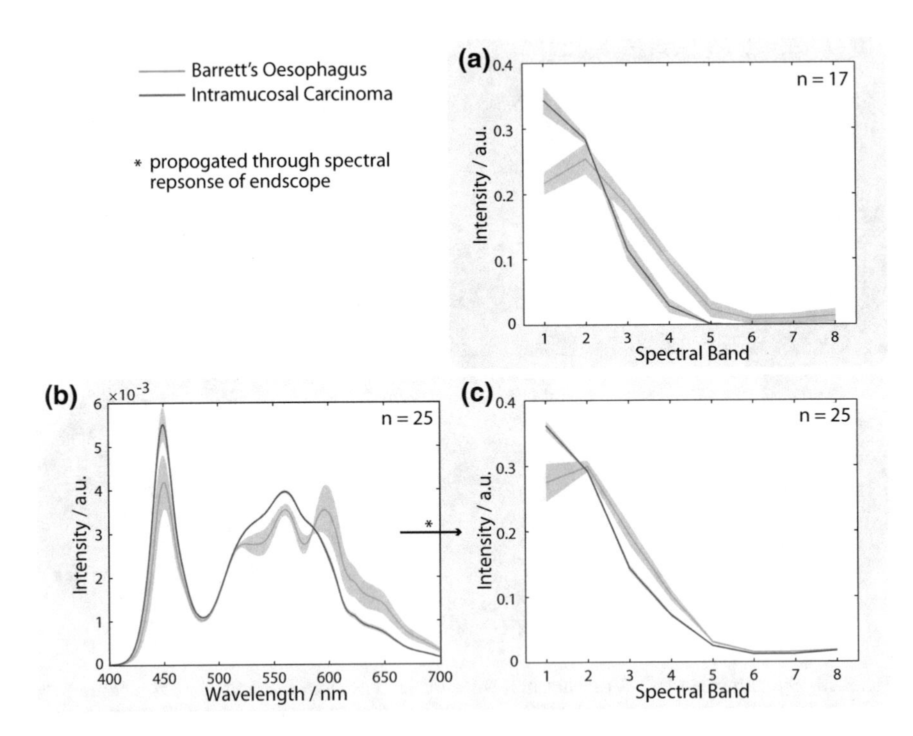

Fig. 5.14 Spectra of Barrett's oesophagus and intramucosal carcinoma measured with the multispectral endoscope. **a** Mean multispectral image spectra from within the ROIs. Shaded areas represent the standard deviation across the ROIs (n = 17 images; 9 Barrett's oesophagus, 8 intramucosal carcinoma). The 8 narrow spectral bands are shown; average FWHM 30 nm; centre wavelengths 553, 587, 629, 665, 714, 749, 791, 829 nm. **b** Mean spectra captured by gold standard spectrometer. **c** The gold standard spectra were propagated through the spectral response of the multispectral endoscope (Sect. 5.5.1.3) to yield modelled 9-band image spectra for comparison with the multispectral image spectra. Shaded areas represent the standard deviation across the spectra (n = 25 spectra; 9 Barrett's oesophagus, 16 intramucosal carcinoma)

realistic clinical scenario simulated using a whole ex vivo porcine oesophagus. The second, designed for multispectral reflectance imaging of endogenous contrast, has been applied in a first-in-human clinical trial, where the measured spectra showed promising differences between Barrett's oesophagus and cancer tissue.

While these results are promising, some challenges remain for further clinical translation of SRDA-based multispectral imaging. The most significant challenge facing SRDA-based multispectral imaging is the limited sensitivity of the device due to the low sensitivity of the SRDA, partly due to the low quantum efficiency (QE) of the underlying sensor (<60%), and partly due to the low transmission of the CFA (~40%). Increased sensitivity is important in order to increase signal for both multispectral reflectance and multispectral fluorescence imaging, but it is particularly important for the latter, where low concentrations of dye must be detected. Furthermore, increased sensitivity enables shorter exposure times, thus allowing frame rate to be increased to achieve video rate imaging (~25 frames per second).

This challenge could be overcome in a number of ways, with increasing complexity. Firstly, and most simply, the illumination power of the system could be increased. This should increase reflectance signal, and fluorescence signal so long as saturation of the fluorophores is avoided. Secondly, should an appropriate collaboration be initiated, the CFA could theoretically be deposited on a higher QE sensor, though this would be expensive. Finally, as a more long-term solution, attenuation associated with imaging though a fibre bundle could be avoided by miniaturising SRDAs and placing them on the tip of the endoscope. Since this requires significant research and development resources, it would likely be executed by one of the major endoscope vendors, and require considerable evidence of the clinical utility of the technique prior to investment.

The large data processing requirements also present a challenge. Currently we perform our spectral unmixing offline, with this taking around 0.025 s per pixel (MacBook Pro, 2.4 GHz Intel Core i5, Memory 8 GB 1600 MHz DDR3), to unmix 4 endmembers, corresponding to around 9 h per 1024 $\times$ 1280 pixel image. Fortunately, since imaging occurs through a fibre bundle, only pixels at the 10,000 fibrelet centres need to be unmixed prior to interpolation, reducing the processing time to 4 min per image. For real time classification, this could be reduced by utilising GPUs, or reducing the number of spectral bands used for unmixing.

Future work should investigate more advanced spectral unmixing techniques. These have been developed for application in a wide range of fields including earth observation, food safety monitoring, pharmaceutical process monitoring, forensics and recently biomedical imaging [2, 37]. Spectral unmixing can be either supervised, where a priori knowledge of endmembers has been used to train the algorithm, or unsupervised, where the algorithm must determine both the endmembers and their abundances based on the variation within the data.

For biomedical spectral unmixing, the endmembers associated with different disease states or tissue types might initially be unknown, hence the need for exploratory pilot studies similar to the MuSE trial presented here. These studies are important, not only because they allow determination of the endmembers in vivo, but because they allow us to determine the intra-patient and inter-patient heterogeneity of these endmembers.

Should the discovered endmembers be well conserved, they can then be used as a priori knowledge for unmixing algorithms. In the simplest case, this means reducing the unmixing problem to an inversion problem aimed at determining the relative abundance of the known endmembers from a measured dataset (e.g. least squares unmixing), but these methods are sensitive to heterogeneity. A more advanced approach is to use a gold-standard dataset, perhaps that collected during the exploratory pilot study, to train a supervised unmixing algorithm (e.g. random forest, neural network). For proper training, a large, well-annotated dataset is required, meaning the images must be clearly outlined and labelled, a time consuming process. However, the larger this training dataset, the more robust the algorithm to inter-patient and intra-patient heterogeneity.

If the endmembers are not conserved between patients, unsupervised unmixing can be used to determine new endmembers in each patient (e.g. k-means clustering, principle component analysis, autoencoders), but this does not supply labels for the segmentation; these must be input by the user.

Ideally, the colour filter array of the SRDA should be optimised for the detection of the differences between spectral signatures of each disease type, whether these be fluorescence or reflectance, potentially increasing sensitivity whilst limiting the number of spectral bands needed to unmix the spectra of interest, thus decreasing processing time. However, producing a one-off custom CFA is prohibitively expensive. In the long term, once the spectral signatures of Barrett's and dysplasia are well validated in single centre clinical trials, and their clinical utility is established, manufacture of multiple custom CFAs for large scale multicentre trials might be more economical, and more likely to gain the significant investment required.

Despite these outstanding challenges, the SRDA-based multispectral endoscope presented here has many advantages: it has received local safety approval for use in humans; it is compatible with insertion through the working channel of standard of care endoscopes, which enabled easy implementation during the first-in-human trials (Translational Characteristic 2); it can be articulated using the familiar standard of care endoscope, which allowed it to be positioned in order to image particular regions of interest in the oesophagus (Translational Characteristic 3). Furthermore, MSI occurs directly on the SRDA, meaning the compact device has fewer optical components, and is more resistant to misalignment than alternative multispectral imaging techniques (Translational Characteristic 2). Additionally, the multispectral data can be used to produce a false colour RGB image to guide the endoscopist (Translational Characteristic 6).

Several steps must be taken to continue the translation of the two multispectral imaging approaches investigated here. For MFI, much of the technical validation in Domain 2 of the OIB Roadmap has been completed (Fig. 1.3): the device has been shown to accurately image fluorophores using a dilution series; the ability to detect fluorophores in a more realistic setting has been demonstrated using ex vivo pig oesophagus; and the device is available, compatible with the clinical environment and approved for use in humans.

Biological validation of the device requires experiments in which the multispectral endoscope is used to image a targeted molecular imaging probe applied to ex vivo human tissue, similar to those described in Sect. 4.6. These experiments biologically validated WGA-IR800 paired with the bimodal endoscope for delineating dysplasia from Barrett's oesophagus. The underlying principles should apply to MFI paired with WGA conjugated to any fluorophore. However, to ensure thorough biological validation, the experiments should be repeated using the multispectral fluorescence endoscope paired with a particular WGA conjugate. Since it is not yet clear when (or which) targeted molecular imaging probes will be manufactured under GMP conditions, and ultimately approved for use in humans, these experiments were deferred to a later date and attention turned to imaging endogenous contrast.

For multispectral reflectance imaging of endogenous tissue contrast, much of the technical and biological validation required in Domain 2 of the OIB Roadmap (Fig. 1.3) is not easily performed using ex vivo tissue. When tissue is excised from the body, endogenous tissue features change irreversibly (Translational Barrier 4). For example, the lack of active blood flow changes the spectrum of tissue by reducing blood oxygenation to 0% and consequently altering the haemoglobin absorption spectrum. Furthermore, tissue autofluorescence can be modified upon exposure to ambient light and tissue structure may be distorted by surgical trauma, or by positioning the tissue on a rigid surface. Ultimately, tissue will degrade unless fixed in formaldehyde or frozen, which further alters properties [38]. The leap between data acquired during ex vivo and in vivo imaging is thus large and data acquired from ex vivo tissues may contain insurmountable artefacts if the tissues are not properly handled, justifying our decision to proceed directly to in vivo multispectral reflectance imaging in a clinical pilot study (MuSE).

In the first trial of this pilot study, MuSE 01, multispectral endoscopy produced promising results. Distinct reflection spectra were observed for regions of Barrett's oesophagus and cancer, suggesting the potential to delineate diseased tissue based on endogenous reflectance using the multispectral endoscope. These spectra were consistent over multiple viewpoints separated by several minutes of imaging, suggesting high repeatability of the approach.

This work also has several limitations. The majority of the spectral differences were observed in the region from 530–650 nm, but since the light source had little power above 700 nm, the reflectance spectrum above 700 nm was not explored. Since our endoscope can image light up to 850 nm (Fig. 5.7), a more broadband light source will be installed for future trials.

The ultimate aim is to delineate dysplasia from Barrett's oesophagus. The majority of the suspicious lesion imaged in MuSE 01 was composed of intestinal metaplasia, with some high-grade dysplasia, but intramucosal cancer was also present. Future trials with exclusively dysplastic lesions are required to evaluate the appearance of dysplasia on MSI, and thus test the potential of the technique to perform our ultimate objective of delineating dysplasia from Barrett's oesophagus.

In MuSE 01, the coregistration challenges described in Chap. 4 were avoided by delineating regions of interest with cautery marks and taking biopsies of the marked regions following multispectral imaging. However, positioning the multispectral endoscope to image the marked regions is difficult, and further trials will be required to practice and improve this. In addition, biopsies from healthy tissue regions were not taken, meaning gold standard diagnosis of these regions relies on the less reliable classification from the endoscopist inspecting the tissue using HD-WLE. In future, biopsies should be taken from the healthy region to confirm the tissue classification with gold standard histopathology.

Nevertheless, considering this was the first trial of the multispectral endoscope in vivo, MuSE 01 went some way to addressing both the primary objectives of our trial (Sect. 5.6.1.1): it was confirmed that the device can be used to capture images in a clinical environment and with minimal interruption to clinical workflow; the appearance of Barrett's oesophagus without dysplasia (but not yet with exclusively dysplasia) using multispectral endoscopy was evaluated, the first step to building a library of diagnostically useful distinguishing characteristics. Given this initial success, a second patient is due to be recruited in late September 2018 (MuSE 02). Over the next 12 months, we aim to recruit 10–20 patients in MuSE, while iteratively making improvements to the device, optimising it to image the emerging OIBs, and to the methodology, optimising coregistration of the in vivo images with histopathology. With this trial underway, and having demonstrated the potential of multispectral imaging for delineation of disease, we proceeded to apply SRDA-based multispectral imaging in rigid endoscopy for the delineation of brain cancer. The development of this system is discussed in the following chapter.

References

1. Lee JH, Wang TD (2016) Molecular endoscopy for targeted imaging in the digestive tract. Lancet Gastroenterol Hepatol 1:147–155
2. Luthman AS, Dumitru S, Quiros-Gonzalez I, Joseph J, Bohndiek SE (2017) Fluorescence hyperspectral imaging (fHSI) using a spectrally resolved detector array. J Biophotonics 10:840–853
3. Luthman S, Waterhouse D, Bollepalli L, Joseph J, Bohndiek S (2017) A multispectral endoscope based on spectrally resolved detector arrays, p 104110A
4. Joshi BP et al (2016) Multimodal endoscope can quantify wide-field fluorescence detection of Barrett's neoplasia. Endoscopy 48
5. Joshi BP et al (2016) Multimodal video colonoscope for targeted wide-field detection of non-polypoid colorectal neoplasia. Gastroenterology 150:1084–1086
6. Yang C, Hou V, Nelson LY, Seibel EJ (2013) Color-matched and fluorescence-labeled esophagus phantom and its applications. J Biomed Optics 18:26020
7. Georgakoudi I et al (2001) Fluorescence, reflectance, and light-scattering spectroscopy for evaluating dysplasia in patients with Barrett's esophagus. Gastroenterology 120:1620–1629
8. Bays R, Wagnie G, Robert D, Braichotte D (1996) Clinical determination of tissue optical properties by endoscopic spatially resolved reflectometry, vol 35
9. Bargo PR et al (2005) In vivo determination of optical properties of normal and tumor tissue with white light reflectance and an empirical light transport model during endoscopy. J Biomed Optics 10:1–15
10. Holmer C et al (2007) Optical properties of adenocarcinoma and squamous cell carcinoma of the gastroesophageal junction. J Biomed Optics 12:014025

11. Thueler P et al (2003) In vivo endoscopic tissue diagnostics based on spectroscopic absorption, scattering, and phase function properties. J Biomed Optics 8:495–503
12. Lu G, Fei B (2014) Medical hyperspectral imaging: a review. J Biomed Optics 19:10901
13. Calin MA, Parasca SV, Savastru D, Manea D (2014) Hyperspectral imaging in the medical field: present and future. Appl Spectrosc Rev 49:435–447
14. Krishnamoorthi R, Iyer PG (2015) Molecular biomarkers added to image-enhanced endoscopic imaging: will they further improve diagnostic accuracy? Best Pract Res Clin Gastroenterol 29:561–573
15. Fawzy Y, Lam S, Zeng H (2015) Rapid multispectral endoscopic imaging system for near real-time mapping of the mucosa blood supply in the lung. Biomed Optics Express 6:2980–2990
16. Kester RT, Bedard N, Gao L, Tkaczyk TS (2011) Real-time snapshot hyperspectral imaging endoscope. J Biomed Optics 16:056005
17. Saito T, Yamaguchi H (2015) Optical imaging of hemoglobin oxygen saturation using a small number of spectral images for endoscopic application. J Biomed Optics 20:126011
18. Johnson WR, Wilson DW, Fink W, Humayun M, Bearman G (2007) Snapshot hyperspectral imaging in ophthalmology. J Biomed Optics 12:014036
19. Mori M et al (2014) Intraoperative visualization of cerebral oxygenation using hyperspectral image data: a two-dimensional mapping method. Int J Comput Assist Radiol Surg 9:1059–1072
20. MacKenzie LE, Choudhary TR, McNaught AI, Harvey AR (2016) In vivo oximetry of human bulbar conjunctival and episcleral microvasculature using snapshot multispectral imaging. Exp Eye Res 149:48–58
21. Martinez-Herrera SE et al (2014) Multispectral endoscopy to identify precancerous lesions in Gastric Mucosa, vol 8509, pp 43–51
22. Leavesley SJ et al (2016) Hyperspectral imaging fluorescence excitation scanning for colon cancer detection. J Biomed Optics 21:104003
23. Han Z et al (2016) In vivo use of hyperspectral imaging to develop a noncontact endoscopic diagnosis support system for malignant colorectal tumors. J Biomed Optics 21:016001
24. Kumashiro R et al (2016) An integrated endoscopic system based on optical imaging and hyper spectral data analysis for colorectal cancer detection. Anticancer Res 3932:3925–3932
25. Panasyuk SV et al (2007) Medical hyperspectral imaging to facilitate residual tumor identification during surgery. Cancer Biol Ther 6:439–446
26. Liu Z, Yan J, Zhang D, Li Q-L (2007) Automated tongue segmentation in hyperspectral images for medicine. Appl Opt 46:8328–8334
27. Akbari H, Kosugi Y, Kojima K, Tanaka N (2008) Wavelet-based compression and segmentation of hyperspectral images in surgery. Lecture Notes in Computer Science (including subseries Lecture Notes in Artificial Intelligence and Lecture Notes in Bioinformatics), vol 5128, pp 142–149
28. Hagen N, Kudenov MW (2013) Review of snapshot spectral imaging technologies. Opt Eng 52:090901
29. Gu X et al (2016) Image enhancement based on in vivo hyperspectral gastroscopic images: a case study. J Biomed Optics 21:101412
30. Leitner R et al (2013) Multi-spectral video endoscopy system for the detection of cancerous tissue. Pattern Recogn Lett 34:85–93
31. Yang C et al (2014) Scanning fiber endoscope with multiple fluorescence-reflectance imaging channels for guiding biopsy 89360R
32. Lee CM, Engelbrecht CJ, Soper TD, Helmchen F, Seibel EJ (2010) Scanning fiber endoscopy with highly flexible, 1 mm catheterscopes for wide-field, full-color imaging. J Biophotonics 3:385–407
33. Tate TH, Keenan M, Black J, Utzinger U, Barton JK (2017) Ultraminiature optical design for multispectral fluorescence imaging endoscopes. J Biomed Optics 22:036013
34. Waterhouse DJ, Luthman AS, Bohndiek SE (2017) Spectral band optimization for multispectral fluorescence imaging, vol 10057, pp 1005709
35. ThermoFisher Fluorescence SpectraViewer

36. Kaye PV et al (2009) Barrett's dysplasia and the Vienna classification: reproducibility, prediction of progression and impact of consensus reporting and p53 immunohistochemistry. Histopathology 54:699–712
37. Bioucas-Dias JM et al (2012) Hyperspectral unmixing overview: geometrical, statistical, and sparse regression-based approaches. IEEE J Sel Top Appl Earth Obs Remote Sens 5:354–379
38. Shim MG, Wilson BC (1996) The effects of ex vivo handling procedures on the near-infrared Raman spectra of normal mammalian tissues. Photochem Photobiol 63:662–671

Chapter 6
Rigid Endoscopy for Intraoperative Imaging of Pituitary Adenoma

In the previous chapters, optical imaging was discussed in the context of upper endoscopy for the surveillance of Barrett's oesophagus, where a wide range of light-tissue interactions (Fig. 1.1) have been utilised and investigated thoroughly in many novel optical imaging techniques, some of which have been deployed in vivo. In contrast, for intraoperative pituitary imaging the majority of light-tissue interactions are yet to be investigated, leaving many potential contrast mechanisms unexploited.

6.1 Pituitary Adenoma

The pituitary gland is a pea-sized organ situated behind the ridge of the nose, attached to the base of the brain by a thin stalk. It is a key component of the endocrine system, responsible for hormonal control of other glands as well as many aspects of normal functioning including growth and blood pressure. Pituitary adenoma, the development of a benign tumour, is the most common disease associated with the pituitary gland, with a prevalence of approximately 17% in the US population, although the majority cause no symptoms [1]. These are a diverse group of tumours, which may be functional or non-functional [1]: functional adenomas usually present with clinical symptoms specific to the increased hormone secretion, while non-functional adenomas present due to mass effect resulting in visual loss due to compression of the optic chiasm or identification incidentally on imaging for other indications [2].

6.1.1 Standard of Care

Transsphenoidal surgical resection is the primary treatment method for pituitary adenomas (Fig. 6.1) [3]. A surgeon inserts rigid endoscopic devices through the nostrils and to the back of the nasal cavity where small openings are made in three bones, the nasal septum, sphenoid sinus and the sella, to reach the pituitary gland.

D. J. Waterhouse, *Novel Optical Endoscopes for Early Cancer Diagnosis and Therapy*, Springer Theses, https://doi.org/10.1007/978-3-030-21481-4_6

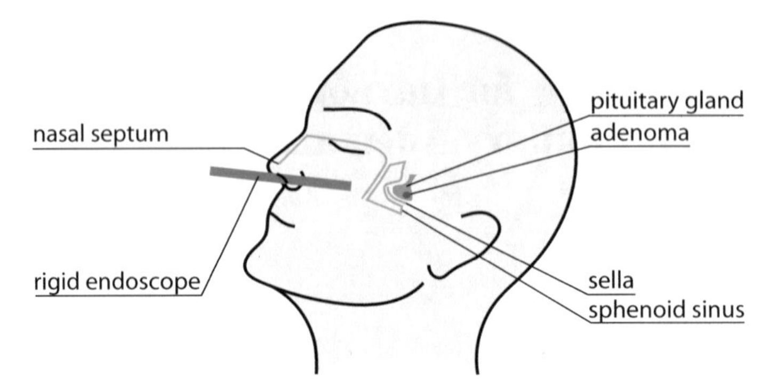

Fig. 6.1 Simple schematic showing transsphenoidal surgery for pituitary adenoma

The surgeon then removes the adenoma guided by white light imaging followed by closing of the sella. It is often difficult to determine the difference between normal pituitary gland, which must be preserved to minimise loss of function, and the pituitary tumour.

Following transsphenoidal surgery, cure rate for acromegaly related adenomas is reported between 46–79% depending on surgical experience [4]. Remission for Cushing's disease related adenomas following transsphenoidal surgery is reported at 89% [5]. Reoperation is difficult: scar formation may distort anatomy and impair adequate resection of tumours. This increases the risk of complications such as loss of normal pituitary function and carotid artery rupture [6]. Furthermore, recurrent tumours are often more rigid [7], and landmarks may be distorted from preoperative radiotherapy or other treatments [8]. Hence, a delicate balance must be struck between maximising completeness of resection and preserving endocrine function from the normal pituitary, so delineation of normal pituitary from adenoma is paramount [9]. Hence there is an unmet clinical need for novel imaging techniques capable of assisting the surgeon in delineating normal pituitary tissue from adenoma.

6.1.2 Advanced Imaging of Pituitary Adenoma

Currently, surgeons rely on pre-operative MRI in order to try and determine where the normal pituitary gland might be in relation to the tumour, for example, if the pituitary stalk is displaced to the right, it is likely that the normal pituitary gland will also be displaced to the right. Following resection, post-operative MRI is also used as the gold standard for assessing the extent of resection. However, the images may contain artefacts from hematoma or surgical packing, so postoperative MRI often takes place a few months after the surgery, when these artefacts are no longer present [10], too late to inform changes to the initial surgery.

MRI can also occur during surgery. Intra-operative MRI (iMRI) offers immediate feedback to surgeons following resection, allowing them to re-examine the site of any residual tumour that shows up on the MRI images. However, it is expensive, potentially requiring extensive remodelling of operating theatres to accommodate the MRI scanner and it prolongs surgery [10]. Furthermore, image artefacts can lead to false positives, so iMRI is not recommended by the Congress of Neurological Surgeons [3]. In recent years, there has been much research into intraoperative optical imaging technologies for transsphenoidal surgery, including narrow band imaging (NBI) [11], 4K imaging [12], ICG fluorescence imaging [13, 14] and optical molecular imaging (OMI) with a fluorescent folate analogue [15].

6.1.2.1 4K NBI

Narrow band imaging (NBI) was described in detail in Sect. 2.2.2.2. On standard NBI, the pituitary gland appears as an arabesque pattern due to the rich vasculature, whereas the tumour has no particular pattern of enhancement [11]. A recently released 4K ultra-high definition (UHD) NBI endoscope (VISERA, Olympus, Japan) allows 3840 × 2160 pixel UHD imaging with NBI, allowing high resolution imaging of the microvasculature [12]. with a short learning curve, since it is similar to familiar standard NBI (Translational Characteristic 3, Translational Characteristic 4). However, the system is bulky (Translational Characteristic 2), using a 55 in. monitor, and the colour balance must be delicately tuned to allow sustained visualisation during bleeding. It remains to be seen whether UHD NBI will make any lasting impact to the SOC.

6.1.2.2 ICG Fluorescence Imaging

Indocyanine green (ICG) is an intravenously administered green fluorescent dye which binds to blood plasma albumin allowing it to be used for vascular imaging. It has a short half-life, an acceptable safety profile and fluoresces in the NIR, allowing good depth penetration (up to a few mm). There is lower vasculature density in adenomas compared to the healthy anterior pituitary [16, 17], so several groups have performed ICG fluorescence endoscopy in transsphenoidal surgery for benign pituitary lesions [13, 14]. ICG is injected and allowed to circulate before imaging, this timing being crucial; the internal carotid artery reaches peak fluorescence first, followed by the intercavernous sinus and then the pituitary itself [18].

In work by Litvack et al. there was higher fluorescence in the pituitary gland than in the tumour (n = 12 patients), but this pilot study was limited by their standard definition endoscope and a lack of quantitative objective results [14]. Their custom endoscope was developed by Karl Storz and a second-generation device is in development. Sandow et al. found that adenoma could be detected, either by lower ICG intensity compared to surrounding tissue, or higher intensity due to uptake [13] with the type of contrast associated with the clinical symptoms of the patient (n = 22 patients) [13]. This variation of ICG signal characteristics makes ICG based resection difficult due to lack of specificity [13]. In contrast to flexible NIR fluorescence endoscopes (Chap. 4), rigid NIR fluorescence endoscopes are commercially available, these being intended for imaging of ICG, but further development is needed since these are currently ≥5 mm in diameter compared to 4 mm standard endoscopes, and neither side facing nor tip-washing devices are available (Translational Characteristic 2) [18, 19].

6.1.2.3 Optical Molecular Imaging

Intraoperative optical molecular imaging of folate receptors has been proven feasible and beneficial in ovarian cancer (n = 10 patients [20], n = 12 patients [21]). Folate receptor alpha (FRα) overexpression has also been reported in non-functioning pituitary adenomas [22–24]. Using OTL38, a fluorescent folate analogue, and a commercially available NIR fluorescence endoscope (Iridium, Visionsense, USA), Lee et al. improved true positive rate and true negative rate for detecting pituitary adenoma in margin specimens from 80 to 89% using white light alone, to 86 and 89% respectively, but many adenomas did not overexpress FRα (no functioning adenomas and half of the non-functioning adenomas overexpressed FRα), a significant limitation [15]. Their future work will focus only on non-functional adenomas since none of the functioning adenomas overexpressed FRα [15].

6.1.2.4 Multispectral Imaging

ICG fluorescence and optical molecular imaging face translational challenges associated with the use of exogenous contrast (Translational Characteristic 1). As mentioned previously, multispectral imaging has the potential to use spatially resolved spectral data to delineate disease based on endogenous contrast [25, 26]. Furthermore, additional spectral information may be used to correct for working distance [27, 28], which was a critical confounder for optical molecular imaging with OTL38 in the work by Lee et al. [15].

MSI has not yet been used in pituitary imaging but liquid crystal tuneable filter (LCTF) based MSI has been used to capture reflectance and fluorescence image cubes in human glioblastoma, an aggressive brain cancer [29]. Using the reflectance signal at 465–485 and 625–645 nm, the fluorescence signal was corrected to yield quantitative fluorescence images displaying intrinsic fluorescence independent of optical properties of the tissue [29].

SRDAs (Sect. 5.3) provide a more robust, compact, low-cost and fast, alternative to LCTFs, providing a promising solution for multispectral endoscopic imaging (Translational Characteristic 2). However, it is currently unclear whether multispectral imaging can be used to delineate healthy pituitary tissue from pituitary adenoma, so in order to evaluate the appearance of pituitary adenoma and healthy pituitary tissue on multispectral endoscopy, a rigid multispectral endoscope was built to capture in vivo data in a pilot clinical study. Here, the design and development of this clinically translatable SRDA-based rigid multispectral endoscope is presented. Although clinical measurement has not yet been achieved, the initial technical characterisation is described and the challenges that must be addressed in order to achieve clinical translation are discussed.

6.2 Endoscope Design

In flexible endoscopy (Chaps. 3–5), light is carried to the back end optics using long flexible fibres, so the back end optics can easily be spatially isolated from the patient, allowing great flexibility in their design. In transsphenoidal endoscopy, the back-end optics are clipped directly onto the rigid endoscope and handheld by the surgeon, so they must be compact and robust (Translational Characteristic 2). We thus designed a system based around a compact SRDA (CMS-V, SILIOS, France) (Fig. 6.2).

Briefly, the system consists of a 4 mm 0° endoscope (Sharpsite® AC Autoclavable Endoscope, Medtronics, USA) coupled to an SRDA (CMS-V, SILIOS, France) using a zoom coupler (18–35 mm zoom coupler, RVA Synergies, UK). The SRDA consists of 9 spectral filters (8 narrow bands; average FWHM 30 nm; centre wavelengths 553, 587, 629, 665, 714, 749, 791, 829 nm; 1 broad band; 500–850 nm), deposited as a 3 × 3 super-pixel across a CMOS sensor (NIR Ruby sensor, UI1242LE-NIR,

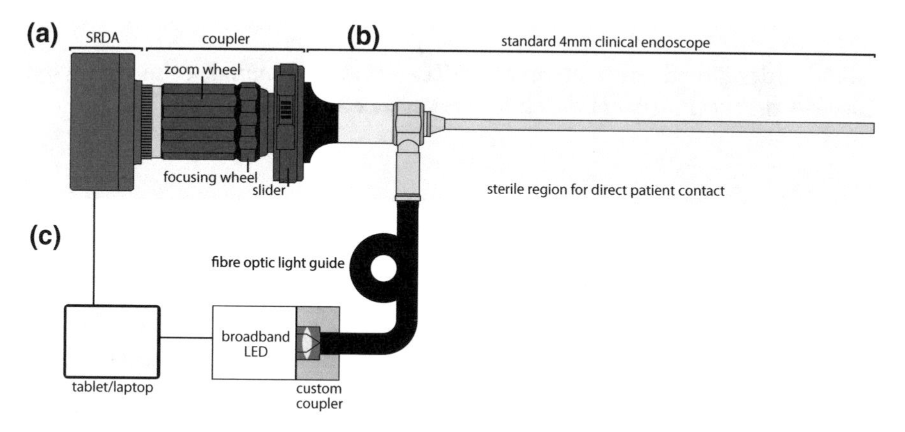

Fig. 6.2 Schematic of the SRDA-based rigid multispectral endoscope. **a** The back end optics are based around a compact and robust SRDA (CMS-V, SILIOS, France) coupled to the rigid endoscope using a commercially available camera coupler (18–35 mm zoom coupler, RVA Synergies, UK). They are easily clipped on and off of the rigid endoscope using a slider, ensuring minimal interruption to the standard clinical workflow. To ensure they remain clean during surgery, they are covered with a plastic bag. This is standard procedure for rigid endoscopy cameras. **b** The region in direct patient contact. **c** Illumination is provided by a broadband ultra-high power LED (T7359, Prizmatix, Israel) coupled to the endoscope via a custom coupler and a fibre optic light cable. The illumination is spatially isolated from the patient as it is passed through a flexible light guide

IDS, square pixel size 5.3 μm). Illumination is provided by a broadband ultra-high power LED (T7359, Prizmatix, Israel) coupled to the endoscope using an achromatic doublet lens (AC254-030-A, Thorlabs, Germany) housed inside a custom coupler with a smooth bore for a fibre optic light guide (495 NA, Karl Storz, Germany). Images were captured in uEye Cockpit (IDS, Germany) and saved as 8-bit BMP files. Data analysis was carried out using Matlab® (MathWorks, USA).

Similar to the flexible endoscopes presented in Chaps. 3–5, this device architecture ensures that no modification has been made to any part of the endoscope that is intended to be in direct patient contact (Fig. 6.2b), meaning approval for use in humans can be granted locally by clinical engineering (Translational Characteristic 2). The back end optics are easily clipped on and off the rigid endoscope using a slider, ensuring minimal interruption to standard clinical workflow (Translational Characteristic 2). Since they are adjacent to the patient, the back end optics are covered with a plastic bag to keep them clean during surgery, as is standard protocol for rigid endoscopy cameras. The illumination is spatially isolated from the patient as it is supplied through a flexible light guide (Fig. 6.2c).

The multispectral data can be used to produce a false colour RGB image familiar to the surgeon, such that the surgeon can continue to survey the patient during multispectral imaging, watching for adverse events such as bleeding, thus ensuring the patient's safety is not compromised during this time (Translational Characteristic 6).

6.3 Technical Characterisation

Following design and construction of the SRDA-based rigid multispectral endoscope, technical characterisation of the system commenced.

6.3.1 Methods

6.3.1.1 Resolution

To determine the limiting resolution of the SRDA-based rigid multispectral endoscope, images of a 1951 USAF resolution test target (#53-714, Edmund Optics, USA) were captured at 6 working distances (7.5–20.0 mm) using external illumination from a broadband halogen light source (OSL2B2, Thorlabs, Germany) to reduce specular reflections. The raw images were simply demosaicked as described in Sect. 3.3.4 and the broad band (500–850 nm) channel images were analysed.

6.3.1.2 Field of View

In order to measure the field of view (FOV), images of a 1 mm checkerboard printed on white paper were captured at 17 working distances (WD), 4–20 mm (error ± 0.1 mm). The resulting images are expected to show barrel distortion defined by:

$$r_u = Ar_d\left(1 + kr_d^2\right) \tag{6.1}$$

where r_u is the radial distance from the center of the ground truth image to a given vertex i in mm, r_d is the radial distance from the centre of the distorted image to the same vertex i in pixels, k is a constant that describes the magnitude of the distortion and A is a constant used to convert between units of pixels and mm.

For each of the images acquired the position of the checkerboard vertices were identified using the built in Matlab function 'detectCheckerboardPoints'. For each of these points, the radial distance to the centre of the image, r_d (in pixels), and the true distance to the centre of the image, r_u (in mm), which is known from the checkerboard pattern, were found.

6.3.2 Results

6.3.2.1 Resolution

Resolution was determined by taking images of a USAF test target. The raw images were demosaicked and the broad band (500–850 nm) channel images were analysed. The Michelson contrast (Eq. 3.17) was calculated for each element and the results plotted against the reciprocal of the line width of the element (Fig. 6.3). A contrast threshold of 1% has previously been reported to be applicable across a wide range of targets and conditions [30], but a threshold of 5% was chosen to avoid effects arising from noise at very low contrast. By finding the intersect of this threshold with exponential fits applied to the data at each WD, the resolution of the endoscope was determined to be 68 ± 7, 83 ± 7, 103 ± 9, 110 ± 10, 130 ± 10 and 150 ± 10 μm, at working distances of 7.5, 10.0, 12.5, 15.0, 17.5 and 20 mm respectively.

6.3.2.2 Field of View

Example images displaying barrel distortion can be seen in Fig. 6.4. The distortion constant k and the constant A were determined by fitting Eq. 6.1 to the data ($R^2 = 0.9773 - 0.9927$). The values of k and A were then used to determine the FOV radius ($=r_u$) based on the radius of the images in pixels ($=r_d$). Combining these data for the 20 working distances, the angle of the FOV was determined to be $96.6 \pm 0.6°$, which compares favourably to the manufacturer specified angle of 102°.

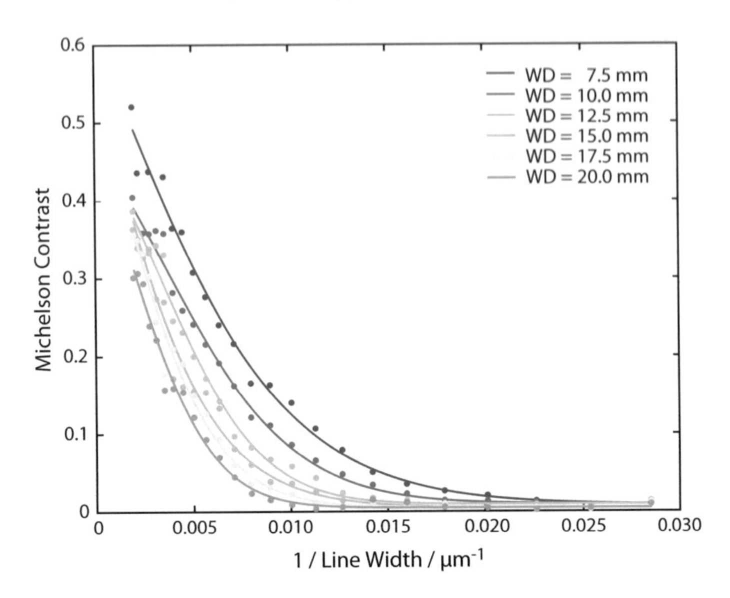

Fig. 6.3 Characterisation of the resolution of the rigid multispectral endoscope. Resolution was characterised by imaging a USAF chart at six different working distances. The resolution was determined as the point where an exponential fit drops below 5% Michelson contrast. $R^2 = 0.9913$, 0.9909, 0.9896, 0.9785, 0.9862 and 0.9886 for working distances of 7.5, 10.0, 12.5, 15.0, 17.5 and 20 mm respectively

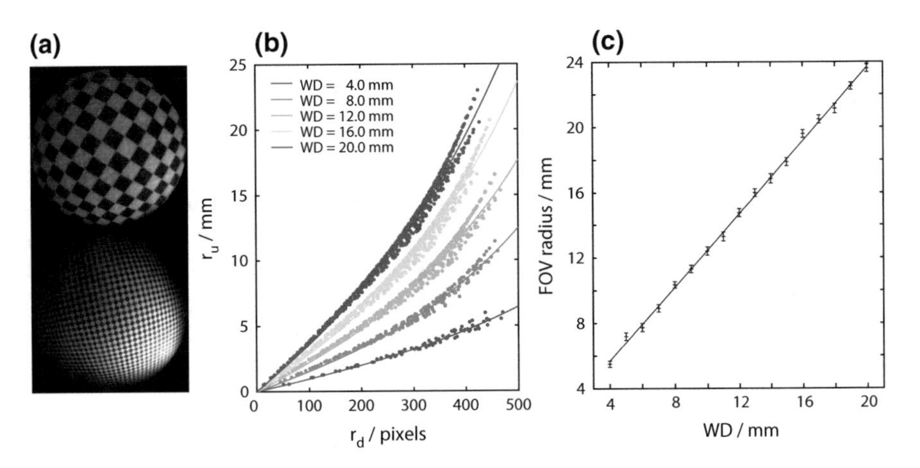

Fig. 6.4 Characterisation of the FOV of the rigid multispectral endoscope. **a** Images of 1 mm checkerboard paper at WD = 4 and 20 mm show barrel distortion. From such an image, r_u and r_d can be measured for several vertices on the paper. **b** Examples of the fit to Eq. 6.1 for images taken at a working distances (WD) of 4, 8, 12, 16 and 20 mm ($R^2 = 0.9799 - 0.9908$). The fit was used to extract the constant A and the distortion parameter k. The values of A and k can be used with Eq. 6.1 to determine the FOV radius ($=r_u$) based on the radius of the images in pixels ($=r_d$). **c** Determined FOV for 17 WDs ($R^2 = 0.9986$). Error bars represent the standard error of the FOV radius derived from the standard errors of the fit parameters A and k. From the fitted line, the angular FOV was calculated to be $96.6 \pm 0.6°$

6.4 Clinical Trial

Having developed a clinically translatable rigid multispectral endoscope, and having performed preliminary technical characterisation, a pilot cohort study was designed, entitled: "Prospective pilot cohort study to assess feasibility of multispectral endoscopic imaging for delineation of the pituitary gland during transsphenoidal surgery for pituitary adenomas" or "Multispectral imaging of adenoma in pituitary surgery (MAPS)" for short. The aim is to recruit 20 patients for investigation in this study, with multispectral imaging occurring as part of the subjects' clinically indicated surgery.

6.4.1 Methods

6.4.1.1 Trial Objectives

The trial is designed to address one primary objective and one secondary objective:

(i) Evaluate the appearance of healthy pituitary tissue and pituitary adenoma at multispectral endoscopy to build a library of useful distinguishing characteristics since these are currently unknown and not easily investigated using ex vivo tissue.
(ii) Assess the feasibility of imaging with the novel multispectral endoscope in endoscopic transsphenoidal pituitary surgery.

6.4.1.2 Trial Design

(i) Potential study participants will be approached in clinic after they have agreed to pituitary surgery for their pituitary tumours. The study will be explained and patient information given with time for reflection before consent is taken to enter the study.
(ii) On the day of surgery, nothing will be changed in terms of patient preparation for their operation. Their pituitary surgery will proceed with endoscopic resection of their pituitary tumour using a standard two surgeon four handed technique. During tumour resection (a. before the pituitary is seen, b. when the pituitary gland is thought to be seen and c. when the tumour resection has finished) the surgeon will take a still picture with the standard white light endoscope for orientation, before removing the standard camera from the end of the endoscope and replacing this with the multispectral camera for two minutes in order to capture intraoperative images.
(iii) As is standard practice during endoscopic pituitary surgery, the tumour will be sent for histopathological analysis.

Images captured with the multispectral endoscope will be used to generate false colour images by assigning the narrow bands centred at 629, 587 and 553 nm to RGB channels respectively. Following completion of the surgery, the surgeon will annotate these images with regions of interest where they are certain of the presence of particular tissue types such as healthy pituitary gland and pituitary adenoma. These regions of interest will be used to determine the multispectral signatures of the different tissue types.

6.4.2 Results

The trial was approved on 17th August 2018. A first patient is due to be recruited in late October 2018 (MAPS 01).

6.5 Conclusions and Future Work

To address the clinical need for better visualisation of adenoma in pituitary surgery, an SRDA-based rigid multispectral endoscope capable of acquisition of 9-band multispectral images in vivo, with a resolution of 68 ± 7 μm, and a FOV of $96.6 \pm 0.6°$ was developed. Rigid SRDA-based multispectral endoscopy shares many of the same challenges of flexible SRDA-based multispectral endoscopy discussed in Chap. 5. SRDA-based multispectral imaging is limited in sensitivity and it remains to be seen whether this will hinder the use of the device in vivo during pituitary surgery. The data processing challenges discussed previously remain here. The rigid endoscope does not introduce a comb artefact as seen in flexible endoscopy. Whilst this affords a greater resolution (83 ± 7 vs. 240 ± 20 μm at WD = 1 cm for the flexible and rigid SRDA-based multispectral endoscopes respectively), it comes at the price of increased processing time, since every pixel must be spectrally unmixed (not just those at fibre centres). For future real-time unmixing, shorter processing times might be achieved by utilising GPUs, or reducing the number of spectral bands.

Despite these challenges, the SRDA-based rigid multispectral endoscope has several advantages: it has received local safety approval for use in humans; the back end optics are easily clipped on and off a rigid endoscope using a slider, ensuring minimal interruption to standard clinical workflow (Translational Characteristic 2); it can be operated in the same way as familiar standard of care endoscopes (Translational Characteristic 3) and the multispectral data can be used to produce a false colour RGB image more familiar to the surgeon, ensuring that the patient's safety is not compromised during multispectral imaging (Translational Characteristic 6). Furthermore, MSI occurs directly on the SRDA, meaning the compact device has fewer optical components, and is more resistant to misalignment than alternative multispectral imaging techniques, which is especially important when the device must be handheld as is necessary in pituitary surgery (Translational Characteristic 2).

Having developed this device, we performed preliminary technical characterisation. This work represents the first steps in Domain 2 of the OIB Roadmap (Fig. 1.3). For further translation, detailed technical validation should be carried out to ensure precision of the technique, and detailed biological validation should be performed, to relate images to underlying pathology. It is not possible to perform these experiments prior to understanding the specific spectral signatures that will ultimately be of interest, and assessment of these is not possible using ex vivo tissue for the reasons discussed previously (Sect. 5.7) (Translational Barrier 4). Therefore, a clinical pilot study (MAPS) to investigate the appearance of pituitary adenoma and healthy pituitary tissue on multispectral imaging was planned.

The first patient is due to be recruited in late October 2018 (MAPS 01). Over the next 12 months, we aim to recruit 10–20 patients in MAPS, iteratively improving the device, methodology and analysis as we proceed. The captured images will be used to determine the spectral signatures of different tissue types, particularly healthy tissue and pituitary adenoma. The device will be modified to optimise precise and repeatable imaging of these spectra. Finally, using these as endmembers, we plan to unmix the multispectral images, to produce maps of the surgical field where different tissue types are clearly delineated. If this is successfully achieved, the next step would be to progress along the OIB Roadmap, carrying out a second clinical trial to assess the clinical utility of these maps.

References

1. Ezzat S et al (2004) The prevalence of pituitary adenomas: a systematic review. Cancer 101:613–619
2. Theodros D, Patel M, Ruzevick J, Lim M, Bettegowda C (2015) Pituitary adenomas: historical perspective, surgical management and future directions. CNS Oncol 4:411–429
3. Kuo JS et al (2016) Congress of neurological surgeons systematic review and evidence-based guideline on surgical techniques and technologies for the management of patients with non-functioning pituitary adenomas. Neurosurgery 79:E536–E538
4. Hazer DB et al (2013) Treatment of acromegaly by endoscopic transsphenoidal surgery: surgical experience in 214 cases and cure rates according to current consensus criteria. J Neurosurg 119:1467–1477
5. Acebes JJ, Martino J, Masuet C, Montanya E, Soler J (2007) Early post-operative ACTH and cortisol as predictors of remission in cushing's disease. Acta Neurochir 149:471–477
6. Bao X et al (2016) Extended transsphenoidal approach for pituitary adenomas invading the cavernous sinus using multiple complementary techniques. Pituitary 19:1–10
7. Berkmann S, Schlaffer S, Buchfelder M (2013) Tumor shrinkage after transsphenoidal surgery for nonfunctioning pituitary adenoma. J Neurosurg 119:1447–1452
8. Esquenazi Y et al (2017) Endoscopic endonasal versus microscopic transsphenoidal surgery for recurrent and/or residual pituitary adenomas. World Neurosurg 101:186–195
9. Verstegen MJT et al (2016) Intraoperative identification of a normal pituitary gland and an adenoma using near-infrared fluorescence imaging and low-dose indocyanine green. Oper Neurosurg 12:260–267

10. Buchfelder M, Schlaffer SM (2012) Intraoperative magnetic resonance imaging during surgery for pituitary adenomas: pros and cons. Endocrine 42:483–495
11. Akutsu N, Taniguchi M, Kohmura E (2016) Visualization of the normal pituitary gland during the endoscopic endonasal removal of pituitary adenoma by narrow band imaging. Acta Neurochir 158:1977–1981
12. Rigante M et al (2017) Preliminary experience with 4K ultra-high definition endoscope: analysis of pros and cons in skull base surgery. Acta Otorhinolaryngol Ital 237–241
13. Sandow N, Klene W, Elbelt U, Strasburger CJ, Vajkoczy P (2015) Intraoperative indocyanine green videoangiography for identification of pituitary adenomas using a microscopic transsphenoidal approach. Pituitary 18:613–620
14. Litvack ZN, Zada G, Laws ER (2012) Indocyanine green fluorescence endoscopy for visual differentiation of pituitary tumor from surrounding structures. J Neurosurg 116:935–941
15. Lee JYK et al (2017) Folate receptor overexpression can be visualized in real time during pituitary adenoma endoscopic transsphenoidal surgery with near-infrared imaging. J Neurosurg 1–14
16. Yamada S, Takada K (2003) Angiogenesis in pituitary adenomas. Microsc Res Tech 60:236–243
17. Jugenburg M, Kovacs K, Stefaneanu L, Scheithauer BW (1995) Vasculature in nontumorous hypophyses, pituitary adenomas, and carcinomas: a quantitative morphologic study. Endocr Pathol 6:115–124
18. Hide T, Yano S, Shinojima N, Kuratsu J (2015) Usefulness of the indocyanine green fluorescence endoscope in endonasal transsphenoidal surgery. J Neurosurg 122:1185–1192
19. Highlights | KARL STORZ endoskope | United Kingdom. Available at https://www.karlstorz.com/gb/en/highlights-tp.htm. Accessed on 14 Sept 2018
20. van Dam GM et al (2011) Intraoperative tumor-specific fluorescence imaging in ovarian cancer by folate receptor-α targeting: first in-human results. Nat Med 17:1315–1319
21. Hoogstins CES et al (2016) A novel tumor-specific agent for intraoperative near-infrared fluorescence imaging: a translational study in healthy volunteers and patients with ovarian cancer. Clin Cancer Res 22:2929–2938
22. Evans C et al (2003) Differential expression of folate receptor in pituitary adenomas. Cancer Res 4218–4224
23. Larysz D et al (2012) Expression of genes FOLR1, BAG1 and LAPTM4B in functioning and non-functioning pituitary adenomas. Folia Neuropathol 50:277–286
24. Evans C, Yao C, LaBorde D, Oyesiku NM (2008) Chapter 8 folate receptor expression in pituitary adenomas: cellular and molecular analysis. Vitam Horm 79: 235–266
25. Lu G, Fei B (2014) Medical hyperspectral imaging: a review. J Biomed Opt 19:10901
26. Calin MA, Parasca SV, Savastru D, Manea D (2014) Hyperspectral imaging in the medical field: present and future. Appl Spectrosc Rev 49:435–447
27. Joshi BP et al (2016) Multimodal endoscope can quantify wide-field fluorescence detection of Barrett's neoplasia. Endoscopy 48
28. Yang C, Hou V, Nelson LY, Seibel EJ (2013) Color-matched and fluorescence-labeled esophagus phantom and its applications. J Biomed Opt 18:26020
29. Valdés PA et al (2012) Quantitative, spectrally-resolved intraoperative fluorescence imaging. Sci Rep 2:798
30. Pelli DG, Bex P (2013) Measuring contrast sensitivity. Vision Res 90:10–14

Chapter 7
Conclusions and Outlook

This thesis establishes three optical imaging techniques based on two endoscopic device architectures to address unmet clinical needs in two indications: detection of dysplasia in endoscopic surveillance of Barrett's oesophagus; and delineation of pituitary adenomas in transphenoidal endoscopic surgery. In establishing these techniques, close attention was paid to the translational characteristics (Chap. 1) to ensure these techniques were appropriate for translation. The first, a flexible endoscope for oesophageal imaging (Chap. 3), was paired with NIR fluorescence optical molecular imaging (Chap. 4) and later SRDA-based multispectral imaging (Chap. 5) for detection of dysplasia in Barrett's oesophagus. The second, a rigid endoscope for intraoperative imaging during transphenoidal surgery, was paired with SRDA-based multispectral imaging for intraoperative delineation of pituitary adenoma (Chap. 6). In addition to the design of the devices, and the development of software for their control, correction methods were established to remove the comb structure associated with the fibre bundle of the flexible endoscopes, and to demosaic the images acquired with SRDAs (Chap. 3). Each of these three techniques was validated and compared with gold standards on measurements across a broad range of biological and non-biological samples. Finally, the two multispectral techniques were translated to in vivo clinical trials.

To ensure swift clinical translation of novel optical imaging techniques, the OIB Roadmap was developed, outlining the steps from discovery of a novel OIB to its widespread implementation in healthcare (Chap. 1). Associated with this, 6 key translational characteristics were defined. Careful consideration of a technique's repeatability and reproducibility, contrast mechanism, instrumentation, operator expertise, co-registration and image reader expertise, can promote its progression along the roadmap. These characteristics influenced many of the decisions in designing the optical imaging techniques presented in this thesis, and ultimately meant that the techniques were suitable for clinical measurements within the 4 year time frame of the work presented in this thesis.

The first part of this thesis described the ongoing work to develop novel optical imaging techniques for detection of dysplasia in patients with Barrett's oesophagus (Chap. 2). For this work, an accessory channel endoscope device architecture was

D. J. Waterhouse, *Novel Optical Endoscopes for Early Cancer Diagnosis and Therapy*, Springer Theses, https://doi.org/10.1007/978-3-030-21481-4_7

chosen to facilitate local safety approval of the devices (Chap. 3). The major drawback of this approach is the comb artefact introduced by the imaging fibre bundle, but this was addressed by the use of decombing algorithms, several of which were compared to ensure that, for a given application, the best decombing method can be chosen. This not only guided decisions for data processing in this work, but it could also be effectively applied across other fibrescopic imaging applications using different weightings with the presented results.

Having identified a translatable device architecture, we designed and validated a bimodal NIR fluorescence and WL reflectance endoscope for OMI of WGA-IR800 to delineate dysplasia in Barrett's oesophagus (Chap. 4). The ex vivo results showed an encouraging correlation between fluorescence signal intensities and histopathological outcome [1–3]. With the majority of the initial validation completed using ex vivo tissue, and the device prepared for application to be locally approved for use in humans, the next step in the OIB Roadmap was to perform first-in-human trials. Much of the groundwork for these trials has been laid, but in order to ensure the quality and safety of WGA-IR800, it must be synthesised under good manufacturing practice (GMP) conditions before it can be used in humans, which is currently prohibitively expensive and therefore, was not possible in the timeframe of the work presented in this thesis.

As the first-in-human trials are yet to take place, it is currently unclear how well this technique will perform in vivo. As discussed in Chap. 4, some instrumentation challenges remain; fluorescence images ideally need to be corrected for variable working distance and the signal to noise ratio will likely need to be improved depending on the final bound concentration of WGA-IR800. The next step will be to apply the technique in vivo to assess the scale of these challenges. Work to find an inexpensive way to synthesise WGA-IR800 in GMP conditions, so that it may be applied in vivo, is ongoing. If this is achieved, a clinical pilot study of OMI using the bimodal endoscope is set to begin.

Due to the demonstrated potential of MSI in cancer detection, an SRDA-based multispectral endoscope was developed (Chap. 5). Initially, this device was applied to multispectral fluorescence imaging, demonstrating the ability to accurately detect fluorescent contrast agents, both in well plates, and in a realistic clinical scenario, using a whole ex vivo porcine oesophagus. The most significant challenge facing SRDA-based multispectral imaging was found to be the limited sensitivity of the device imposed by low sensitivity of the SRDA. This is important for real time imaging, especially in scenarios where signal is low, for example, when imaging low concentrations of fluorophore. Fabrication of custom sensors, for example by depositing the CFA on a higher QE sensor or miniaturising SRDAs and placing them on the tip of the endoscope, have the potential to increase sensitivity, but these require significant investment in research and development. Prior to these sensors being realised, further work on OMI would be likely to deploy the higher sensitivity bimodal system described in Chap. 4. In the meantime, we await the GMP synthesis and approval of a molecular contrast agent for use in humans.

Having already developed and characterised an SRDA-based multispectral endoscope, and given the potential of multispectral imaging to detect endogenous contrast, we modified the multispectral endoscope for reflectance imaging of endogenous reflectance. When tissue is excised from the body, endogenous tissue features change irreversibly (Translational Barrier 4), so technical and biological validation of multispectral reflectance imaging is not easily performed using ex vivo tissue, hence in vivo evaluation is required. Therefore, an exploratory pilot clinical study to assess the feasibility of multispectral endoscopic imaging for detection of dysplasia in patients with Barrett's oesophagus was planned (MuSE).

In the first trial of this pilot study, MuSE 01, multispectral endoscopy produced promising results. Distinct reflection spectra were observed for regions of Barrett's oesophagus and cancer, suggesting the potential to delineate diseased tissue based on endogenous reflectance using the multispectral endoscope. These spectra were demonstrated to be consistent over multiple viewpoints separated by several minutes of imaging, suggesting high repeatability of this approach.

Our work on flexible endoscopy, detailed in Chaps. 2–5, established a framework for swiftly translating novel flexible endoscopic techniques to in vivo clinical pilot studies. It was shown that the PolyScope accessory channel endoscope is a useful device architecture for translating novel contrast mechanisms into clinical trials. Effective decombing algorithms necessary to overcome the comb artefact introduced by the PolyScope were developed and evaluated. A protocol for performing simulated endoscopy that can be executed in the Department of Physics, adjacent to our optics labs, was established using whole ex vivo porcine oesophagus, allowing new techniques to be tested in a realistic setting during the development phase, without the need for upheaval and transport to a clinic, where entry, time and space restrictions may prohibit effective early stage validation. Safety approval for PolyScope-based devices has been granted, this being sufficiently broad to cover alternative back end optics, enabling device modifications to be made without delays associated with seeking new or amended approval. Based on this framework, several novel endoscopic imaging techniques, including a hyperspectral endoscope and a phase sensitive endoscope, will be applied in vivo in the next few years.

Following the promising initial results of SRDA-based multispectral imaging in Barrett's, we aimed to apply this technique to intraoperative imaging during transsphenoidal surgery for resection of pituitary adenoma (Chap. 6). For this work a rigid SRDA-based multispectral endoscope was constructed, preliminary technical characterisation was performed and a pilot clinical study (MAPS) was designed and accepted. The first patient is due to be recruited in late October 2018.

Over the next 12 months, we aim to recruit 10–20 patients in both the MuSE and the MAPS studies, iteratively improving the devices, methodology and analysis as the trials proceed. The captured images will be used to build a database of tissue spectral properties from which the spectral signatures of different tissue types, Barrett's oesophagus versus dysplasia and healthy pituitary versus adenoma, can be determined. Using these spectral signatures as endmembers, images clearly delineating different tissue types could be produced. The next step would be to carry out prospective clinical trials to determine the clinical utility of these techniques, crossing Translational Gap 1 in the OIB Roadmap. Were their clinical utility proven in multicentre clinical trials, these techniques might one day cross Translational Gap 2 to be incorporated into routine healthcare, improving upon the current standard of care.

References

1. Waterhouse DJ et al (2016) Design and validation of a near-infrared fluorescence endoscope for detection of early esophageal malignancy. J Biomed Optics 21:084001
2. Neves AA et al (2018) Detection of early neoplasia in Barrett's esophagus using lectin-based near-infrared imaging: an ex vivo study on human tissue. Endoscopy 50:618–625
3. di Pietro M et al (2016) Detection of dysplasia in Barrett's oesophagus using lectin-based near infra-red molecular imaging: an ex-vivo study on human tissue. In Proceedings of the British Society of gastroenterology meeting (2016)

Printed in the United States
By Bookmasters